Evolutionary Principles

TERTIARY LEVEL BIOLOGY

A series covering selected areas of biology at advanced undergraduate level. While designed specifically for course options at this level within Universities and Polytechnics, the series will be of great value to specialists and research workers in other fields who require a knowledge of the essentials of a subject.

Titles in the series:

Experimentation in Biology	Ridgman
Methods in Experimental Biology	Ralph
Visceral Muscle	Huddart and Hunt
Biological Membranes	Harrison and Lunt
Comparative Immunobiology	Manning and Turner
Water and Plants	Meidner and Sheriff
Biology of Nematodes	Croll and Matthews
An Introduction to Biological Rhythms	Saunders
Biology of Ageing	Lamb
Biology of Reproduction	Hogarth
An Introduction to Marine Science	Meadows and Campbell
Biology of Fresh Waters	Maitland
An Introduction to Developmental Biology	Ede
Physiology of Parasites	Chappell
Neurosecretion	Maddrell and Nordmann
Biology of Communication	Lewis and Gower
Population Genetics	Gale
Structure and Biochemistry of Cell Organelles	Reid
Developmental Microbiology	Peberdy
Genetics of Microbes	Bainbridge
Biological Functions of Carbohydrates	Candy
Endocrinology	Goldsworthy, Robinson and Mordue
The Estuarine Ecosystem	McLusky
Animal Osmoregulation	Rankin and Davenport
Molecular Enzymology	Wharton and Eisenthal
Environmental Microbiology	Grant and Long
The Genetic Basis of Development	Stewart and Hunt
Locomotion of Animals	Alexander
Animal Energetics	Brafield and Llewellyn
Biology of Reptiles	Spellerberg
Biology of Fishes	Bone and Marshall
Mammal Ecology	Delany
Virology of Flowering Plants	Stevens

TERTIARY LEVEL BIOLOGY

Evolutionary Principles

PETER CALOW, B.Sc., Ph.D.

Reader in Zoology
University of Glasgow

Blackie

Glasgow and London

Distributed in the USA by
Chapman and Hall
New York

6/1983
Biol.

Blackie & Son Limited
Bishopbriggs, Glasgow G64 2NZ

Furnival House, 14–18 High Holborn, London WC1V 6BX

Distributed in the USA by
Chapman and Hall
in association with Methuen, Inc.
733 Third Avenue, New York, N.Y. 10017

British Library Cataloguing in Publication Data

Calow, Peter
    Evolutionary principles.—(Tertiary level biology)
    1. Evolution
    I. Title   II. Series
    575.01   QH366.2
    ISBN 0-216-91396-9
    ISBN 0-216-91395-0 Pbk

Library of Congress Cataloging in Publication Data

Calow, Peter.
    Evolutionary principles.

    (Tertiary level biology)
    Bibliography: p.
    Includes index.
    1. Evolution.   I. Title.   II. Series.
    QH366.2.C34   1983   575   82–17834
    ISBN 0-412-00321-X (Chapman & Hall)
    ISBN 0-412-00331-7 (Chapman & Hall: pbk.)

Filmset by Advanced Filmsetters (Glasgow) Ltd

Printed in Great Britain by
Thomson Litho Ltd, East Kilbride, Scotland

Preface

The last few years have seen a number of new books on evolutionary biology. However most of these are either large or specialized. This is an attempt to produce a thin, general version for undergraduate use. Thinness, of course, demands selectivity, and the aim has been to concentrate on the principles of the subject rather than on the details—principles, that is, of both theory and practice. Thinness also sometimes means that a certain level of knowledge is assumed in the readership, but I hope that this is not the case here, and my intention has certainly been to produce something that is as intelligible to the uninitiated as it is to the well-informed. As for the bibliography, I refer, where possible, to reviews rather than primary sources, so a citation should not be taken to imply any sort of precedence.

In developing the theme, I have adopted a loosely historical approach, not only because I believe that this makes for more interesting reading but also because the subject, like the subject it addresses, has evolved under the critical eye of a selective process. Problems have been perceived, hypotheses have been formulated to explain them, facts have been amassed to test the hypotheses, more problems have been perceived, more hypotheses formulated, and so on. I hope that something of this emerges from the book and that, in the end, the reader is left in no doubt about what a scientific approach is and how it has been applied in evolutionary biology.

P.C.

Contents

CHAPTER ONE

HISTORY OF EVOLUTIONARY PRINCIPLES

1.1 Introduction

There are two firm observations that can be made about the living world: first, that it comprises a rich diversity of animal, plant and microbial life, and second, that all living organisms seem to be well-fitted for the problems posed by the environments in which they live. It is conceivable that either the characters of organisms are fixed and have remained so since the origin of life, or that they are mutable and that their diversity and adaptability have unfolded progressively with time. The dynamic process is now termed *evolution*. However, several possible mechanisms have been proposed to account for it, ranging from divine intervention to natural selection. The latter view, initiated by Darwin, is now dominant and it will be the aim of this book to summarize the complex collection of ideas and facts that have become associated with it, particularly over the last hundred years. To do this adequately, however, it is necessary to appreciate the historical and philosophical background from which the Darwinian Revolution emerged, and so this chapter will present a short historical review and some philosophical comments.

1.2 Fixity, design and creation

Nullae speciae novae was the catch-phrase for the early systematists, including Linnaeus (1707–1778). Species were the units of creation as prescribed in Genesis and were therefore immutable. Moreover, the reason they were so well-fitted for the challenges presented by everyday life was that they had been designed by God for specific functions in nature. John Ray (1626), clergyman, naturalist and early systematist, saw the fitness of species as evidence for the existence of a Designer, and this

1

Argument from Design was made even more explicit later by William Paley (1743–1805). Animals and plants are wonderful bits of machinery, they asserted, more wonderful than any man-made machine, and so they must be the product of an intelligence more wonderful than that of Man. (There was a later, more subtle, version of this argument which emphasized harmony of form rather than utility. These are described as the idealist and utilitarian positions, respectively.)

Yet there were biological and even theological and philosophical problems with this creationist position. For example, many fossilized organisms had been discovered that no longer existed on earth, and yet it was inconceivable that the perfect products of an omnipotent designer could ever have become obsolete. Cuvier (1769–1832), a very influential French biologist, offered catastrophes as a way out. A series of upheavals, Noah's Flood being one, had removed some of the species initially created by God. Apparently useless characters were another worry because, again, inferior design could hardly be attributed to a super-intelligent designer. Buffon (1707–1788), another French biologist, wrote of the pig:

> it has evidently useless parts, or rather parts of which it cannot make any use; toes all the bones of which are fully formed and which, nevertheless, are of no service to it.

To explain such useless characters he suggested that the Supreme Being had created perfectly designed types embodied in the original species, but that new species arose from them by a process of hybridization and degeneration. Thus the ass was supposed to be a degenerate horse and the ape a degenerate man. But this meant that the assumption of strict fixity of species had to be relaxed and the concept of *nullae speciae novae* disappeared from the last, revised edition of Linnaeus' taxonomic tomb, the *Systema naturae*. (There were also deeper theological and philosophical flaws in the Argument from Design concerned with the origin of imperfections and the image of the Creator as a humanoid designer, but these will not be considered further here.)

1.3   Programmed evolution

Even some of the enlightened medieval scholars considered the story of creation in Genesis to be a myth. Augustine (353–430) likened the work of the Creator to the progressive growth of a tree, and Aquinas (1225–1274) similarly saw creation as a process whereby the powers given to matter by God progressively unfolded.

It was in this sense that Charles Bonnet (1720–1793) first applied the term evolution (*e-volvo* = to unfold), extrapolating from progressive embryogenesis (in individuals) to the development of species. He and Treviranus (1776–1853), famed as the originator of the term 'biology', were of the opinion that the design of all life-forms was contained as germs in the first being. However, Bonnet conceived of this unfolding as the working out of a rigidly determined programme of change, whereas Treviranus saw evolution as the calling into play of an endless variety of form assumptions (programmed into each organism by the Creator) according to the needs of survival in a changing world. Organism and environment thus interacted to generate all extant and extinct species.

These views of the world, as a giant organism endowed with a mysterious developmental impulse, can be traced back to Plato (427–347 BC) and were carried forward into the eighteenth and even nineteenth century by evolutionists like Goethe (1749–1831), Kant (1724–1804) and Chambers (1802–1871), the so-called Naturphilosophers.

1.4 Less rigidly programmed evolution

> These as successive generations bloom,
> New powers *acquire* and larger limbs assume,
> Whence countless groups of vegetation spring,
> And breathing realms of fin and feet and wing.

Thus wrote Erasmus Darwin (1731–1807)—Charles Darwin's grand-father—in his poem, *Temple of Nature* (London, 1806). In this and his book, *Zoonomia* (London, 1794–1796), he proposed a less rigid system of determined evolution based on the acquisition of new characters by the organisms themselves. Lamarck (1744–1820) formulated a strikingly similar theory in his *Philosophie Zoologique* (Paris, 1809). In fact, Charles always suspected that he had plagiarized the ideas of Erasmus, but that was never certain.

Both theories involved three main principles.

(1) All organisms were supposed to have an *innate power* to progress towards a more complex and perfect form. For Erasmus this 'faculty to improve by its own inherent activity' was God-given, but for Lamarck the '*pouvoir de la vie*' was supposed to emerge out of the organization of living beings. Subtle fluids—caloric and electricity—flowed through the organism, not only maintaining order but establishing more order.

(2) All organisms had an inner disposition (what Lamarck called a *sentiment intérieur*) which caused the performance of actions sufficient to meet the needs (survival) created by a changing environment. The environ-

ment does not cause change, it causes the *need* for change, which is recognized and acted upon by the organism. A mysterious principle did the job for Erasmus, but the subtle fluids did it for Lamarck.

(3) Characters acquired through new use were transmissible from one generation to another by a process which was never clearly defined by either Darwin or Lamarck. Similarly, characters lost through disuse in one generation did not reappear in future generations.

Using these principles, Lamarck explained the evolution of tentacles in the head of a snail in the following way. As the snail crawls along it *finds the need* to touch objects in front of it and *makes efforts* to touch them with the foremost part of its head. In so doing, it sends subtle fluids into these extensions. Ultimately nerves grow into them and the tentacles are made permanent. Moreover, these acquired tentacles are transmitted from parent to offspring; i.e. they are heritable. The reader will probably be familiar with a similar account of the evolution of the giraffe's neck and should be able to appreciate how all adaptations might be accounted for in this way. It is to be noted, however, that the heritability of acquired characters is only a part of the Lamarckian explanation of evolution. The other, more important part concerns *how* the characters are acquired; that is, according to the immediate and perhaps even future needs of survival perceived in some mysterious way by organisms.

1.5   Development and demise of Darwinism

Charles Darwin (1809–1882) had no quarrel with Lamarck—that 'justly celebrate⁴ naturalist', as he refers to him in the historic introduction to the fourth edition of the *Origin of Species*—on his idea that life evolves, or, indeed, on his idea that characters acquired during the life of an organism might be transmissible, in a heredity sense, to its offspring. It was on the mechanism of evolution that Darwin parted company with both Lamarck and his own grandfather. They allowed the intervention of mysterious forces enabling organisms to select for themselves just those characters which caused evolutionary progress to more and more perfect forms of adaptation. But, without an extremely sophisticated and intelligent discriminatory power, could organisms distinguish between useful acquired characters (which are usually rare) and deleterious ones (which may be common, e.g. wounds, broken limbs, the after-effects of illness etc.)? And how can organisms be said to be moving towards perfection without bringing design and the Designer back into focus? What Charles Darwin wanted to do was to formulate an explanation of organic evolution without reference, explicit or implicit, to such subjective and anthropomorphic ideas.

In his book, *Origin of Species* (first published by John Murray, 1859), Darwin replaced the idea of directed variation (brought about by a *sentiment intérieur*) with random variation that was heritable. On the origin of this variation, he was a little vague—it might, he thought, be due to environmental influences on the gonads—and, as we shall see below, he was equally vague about the mechanism of inheritance. What he was perfectly clear about, however, was that no matter how the variation originated, it did so without any reference to the needs of the organism. Moreover, for Darwin the variation upon which evolution was based involved very small continuous changes away from the original form. He conceded that large abrupt changes might occur, but claimed that these were rare and that most led to monstrosities. Indeed this, he suggested, was to be expected, for to believe that a major change in morphology and physiology could (*a*) coadapt with the existing organization of organisms and (*b*) be of selective advantage to them was to 'enter into the realms of miracle and to leave those of Science'. Finally, instead of Lamarck's mysterious striving for perfection, Darwin explained evolution in terms of a completely ordinary and non-mysterious process. Animal and plant breeders, he observed, had, by the selective breeding of variants, been able to effect considerable changes in the form of domestic plants and animals. This was *artificial selection*. Similarly, Darwin thought that the struggle between organisms in an overcrowded world for finite resources would ensure that only the fittest survived and that this would lead to a kind of *natural selection* of variants—an idea stimulated by the writings of Malthus on economics (*An Essay on the Principle of Population*, first published in London, 1798).

The gist, then, of Darwin's theory of the mechanism of evolution was as follows: (1) individuals within species show considerable but continuous variation in form and physiology; (2) this variation arises in a random fashion and is heritable; (3) the potential for increase within populations of animals and plants is considerable; (4) but resources are limited and so individuals in a population struggle for their own existence and that of their offspring; (5) hence only some (what Darwin called the fittest) survive and leave offspring with the same traits; (6) through this *natural selection* of the fittest, species become represented by individuals which are better and better adapted.

Given the premises in this Darwinian argument—variability, heritability and finite resources—the consequences (evolution by natural selection) follow automatically; and some would say too automatically for them to be taken seriously as a scientific hypothesis. It is the fit that survive because, by definition, those that survive are fit, and boiled down to this, Darwinism sounds rather hollow and circular. Yet it is not. First, the effects that traits have on survival and reproduction are not fixed but vary from taxon to taxon and habitat to habitat, and so the fitness value of particular traits invites investigation and explanation (see Chapter 3). Second, 'survival of the fittest' is only a crude summary (borrowed by Darwin from Herbert Spencer, 1820–1903) of what was intended by Darwin. Below the tip of the theoretical iceberg were and are a number of non-obvious assumptions on the source of variation and the mechanism of heredity (see Chapter 2). In principle these might be wrong and are testable, and it is this (i.e. the potential fallibility and testability of the theory) that makes Darwinism interesting from a scientific point of view.

But granted the logical tightness and scientific respectability of Darwinism, is it really credible? Could the selection of random variation be expected to produce the complex and integrated pieces of machinery associated with organisms? Paley had suggested that contemplating an organ like the eye would cure anyone of atheism, for such a complex, highly integrated system as this could not possibly have arisen by chance. Darwin himself pointed out, however, that if a continuous sequence is found to exist in the living world from simple, rudimentary, light-sensitive organs to the eyes we find in vertebrates, each being functional and each being of some benefit to the bearer, then the Paleyan problem disappears, for then it is not necessary to think of the eye forming in one step, but by a series of steps linked by heredity continuum and promoted by natural selection. Yet in private Darwin was not so confident, admitting to one of his colleagues that, though the contemplation of the eye had not been sufficient to effect his conversion, it did often send him into a cold sweat! Even when we have the 'stepping stones', the evolution of a delicately balanced and finely co-ordinated organ by a large number of randomly conceived steps is hard to accept, and many people still hold Paley's view. (Ridley (1982) gives a historical review of the challenge presented by complex coadapted organs for Darwin's theory.)

At least we now have time on our side—Darwin didn't. That is to say, we now know, with a fair degree of certainty, that there has been a span of some 3500 million years since the origin of the first living organism, and since the emergence of the first rudimentary, light-sensitive eye-spots in animals like flatworms, there has been a span of over 1000 million years

Table 1.1 Currently accepted time scale for the history of the earth and the evolution of life

Time*	Eras	Epochs	Major evolutionary events	Time†
1		Pleistocene ⎫	Man	11:59 pm
10		Pliocene ⎬ Neogene		
30	Cenozoic	Miocene ⎭		
40		Oligocene ⎫		
60		Eocene ⎬ Paleogene		
75		Paleocene ⎭		11:00 pm
135		Cretaceous	Mammals	
165	Mesozoic	Jurassic	Flowering plants	
205		Triassic		9:30 pm
230		Permian	Gymnosperms	
250		Pennsylvanian ⎫	Reptiles	
280		Mississippian ⎬ Carboniferous		
325	Paleozoic	Devonian	Amphibians	8:30 pm
			Insects	
360		Silurian	Land plants	8:00 pm
			Fishes	
425		Ordovician	Algae	
500		Cambrian	All major invertebrate phyla	6:00 pm
3000	Precambrian	Proterozoic	First fossils (1000)	1:00 pm
10 000		Archeozoic	First life (3000–4000) Origin of earth (10 000)	12:00 Midnight

* Time in millions of years since beginning of epoch.
† Time on a 24-hour scale since origin of life—for perspective.

(Table 1.1). In these unimaginable epochs of time even improbable events become conceivable. But William Thomson (1824–1907), later Lord Kelvin, clipped the time available for Darwin by claiming to show that the earth could not be more than 100 million years old in total. This miscalculation threw Darwin and the Darwinists into a state of uncertainty, and served to undermine their credibility as far as the other scientists of the day were concerned.

On top of all this Darwin had associated himself with a mechanism of heredity, involving the blending of parental contributions, which was fatal for his small variations. In this, he supposed that contributions from each parent mixed to form offspring of intermediate character. But on this basis, advantageous variants simply could not be selected, for they would ultimately be swamped by disadvantageous ones! Consider what would happen, invited Fleeming Jenkin, one of the reviewers of the *Origin of Species* (*North Brit. Review*, 1867, 149–171), to the inheritable traits of a white man shipwrecked on an island of negroes. No matter how 'superior' they might be (the reviewer's words) it was possible to show by simple mathematics that they would be swamped by mixing with those of the natives within just a few generations.

Darwin's response, both to the problems posed by his blending inheritance and to those raised by Lord Kelvin, was to allow an increasing role for the heritability of characters acquired in the life of the organism in his theory of evolution. In this way, relatively large and integrated changes could be assimilated into the heredity material in a short time. But Darwinism then became less distinct from Lamarckism and it posed the old problem of how some acquired characters and not others became coded into the heredity material. Because of this, Darwinism became even less credible, and in the early part of the century we find Nordenskiöld writing in his classical *History of Biology* (Tudor Publ. Co., 1929) a long chapter chronicling the decline and even demise of Darwinism. 'Darwin's theory of the origin of species,' he wrote, 'was long abandoned.' Apart from a reappraisal of the time available for evolution, what was required was a better understanding of the hereditary mechanism, and this we consider in the next chapter under the heading of Neo-Darwinism. A Neo-Lamarckian camp also lingered on for a short time (but cf. Section 2.2.4) advocating natural selection *with* acquired characters. St George Jackson Mivart, for example, in his *On the Genesis of Species* (Macmillan, 1871) could not see how evolution would proceed by the gradual accumulation of small changes, since most intermediate character states would be maladaptive (see 5.3). He therefore suggested that evolution would proceed in jumps, and that the basis of these was the acquisition of useful characters. More will be said about this in Chapter 5.

1.6   Conclusions

In conclusion, it is worth re-emphasizing the distinction between (*a*) the view that life has evolved and (*b*) theories on how it has evolved. By the time the *Origin of Species* was published, many people were prepared to accept the impressive array of evidence in favour of evolution, in particular

   (1) the existence of fossils suggested that faunas and floras had changed with time;

(2) the existence of fundamental similarities in the morphology, physiology and embryology of different species suggested common ancestry;
(3) the existence of geographical discontinuities in faunas and floras associated with physical barriers (e.g. Australia with its distinctive marsupial mammals, and islands—like the Galapagos group visited and commented upon by Darwin—with distinctive plant and animal species) suggested a natural process of changes in the characters of the species 'trapped' in the isolated areas.

There were, nevertheless, still many possible explanations of evolutionary change. Darwin presented one, natural selection of random variation, which was less anthropomorphic than the others and also supported by a number of observations. In particular

(1) natural populations contain individuals which show considerable and often continuous variation;
(2) breeders are able to effect impressive morphological and physiological changes in domestic animals by the selection of this variation;
(3) adaptations are based on the modification of common themes and are often not as good as they could have been had they been constructed *de novo*, without the constraints imposed by pre-existing structures. Imperfect solutions, apparently useless characters and vestigial organs are, somewhat paradoxically, support for a process of evolution based upon natural selection of random variation within the constraints of pre-existing forms. They do not readily fit in with the concept of an Omnipotent Designer.

1.7 Further reading

A good book on the general history of biology is that of Smith (1976). Comprehensive reviews on the history, philosophy and social impact of Darwinism are those of Gillespie (1979), Moore (1979) and Ruse (1979, 1982). Information on Darwin the person, his work and his voyage on the Beagle can be found in Moorehead (1969), George (1982) and Howard (1982).

CHAPTER TWO

MECHANISMS OF INHERITANCE

2.1 Mendelian mechanism

2.1.1 *Basic principles*

Darwin realized that a clear understanding of inheritance was essential to an appreciation of the evolutionary processes. Yet he got entangled with a theory which worked against natural selection, 'diluting' rather than 'concentrating' advantageous variants. It was Gregor Mendel (1822–1884) who developed a theory of inheritance more compatible with Darwinian evolution, and who laid the foundations of the modern science of genetics. He gave a paper on the results of his breeding experiments to the Brünn Natural History Society in 1865, and this was published in the transactions of that society the following year. However, his work went largely unnoticed until 1900 when Correns (in Germany), de Vries (in Holland) and von Tschermak (in Austria) rediscovered it and recognized its significance and importance for Darwin's theory.

Mendel studied the garden pea (*Pisum sativum*), and by so doing he was able to exploit a number of useful features of these plants in experimenting

Table 2.1 Traits studied by Mendel. Dominants in upper case

Trait	Variants
Seed shape	ROUND and wrinkled
Seed colour	YELLOW and green
Seed coat colour	COLOURED and white
Pod shape	INFLATED and wrinkled
Pod colour	GREEN and yellow
Flower position	AXIAL and terminal
Stem length	LONG and short

with them: (1) seeds and plants were available in a wide array of discretely different forms; (2) left to themselves they could self-pollinate because each flower contained male and female parts; (3) controlled crosses could be effected by clipping the male parts from flowers (to prevent selfing) and using pollen from other plants for fertilization.

Mendel chose to work on several traits (Table 2.1) and by repeated self-pollination (selfing) established pure lines of them. Thus he deduced that characters bred true over many generations and this suggested that the hereditary factors controlling them must be very stable. From subsequent breeding experiments Mendel noted that if two pure breeding types, differing by a pair of characters, were crossed, only one character was represented in the next, or first filial (F_1) generation. However, if these progeny were mated, their progeny (F_2 generation) consisted of some individuals with parental traits and others with grandparental traits that were absent in the F_1 generation. These grandparental traits therefore segregated out in the F_2 generation, and this meant that there could have been no blending of genetic material in the F_1 generation. Finally, Mendel noted that the traits in the F_2 generation were represented in a fixed ratio of approximately 3 of one and 1 of the other.

Mendel's interpretation of these data was that (1) the hereditary determinants (what he called factors and we now call *genes*) are particulate; (2) they do not blend; (3) each adult has two such determinants (what we now call *alleles*) for each character (trait) and these are segregated between gametes—i.e. one per gamete; (4) each adult may contain two identical alleles (we now say is *homozygous*) or two different alleles (*heterozygous*) and in the latter case the expression of one is dominant over the other (see Table 2.1); (5) the union of gametes is random. Point 3 is often referred to as Mendel's first law. It means that heterozygous parents produce equal quantities of gametes containing the contrasting alleles.

Mendel represented the pairs of genes by letters; dominant alleles with upper-case and recessive with lower-case letters. Using this convention the results and interpretations of one of his experiments, crossing round (dominant = R) and wrinkled (recessive = r) seeds, is summarized in Figure 2.1. This precisely predicts a ratio of 3 round to 1 wrinkled seeds in the F_2 generation and the actual ratios obtained by Mendel in his experiments were 5474 round and 1850 wrinkled seeds; i.e. 2.96 round : 1 wrinkled.

Mendel also went on to discover that if two more contrasting characters are present in the grandparents, these also segregate in the F_2 generation in such a way that they do not necessarily stay together. This led Mendel

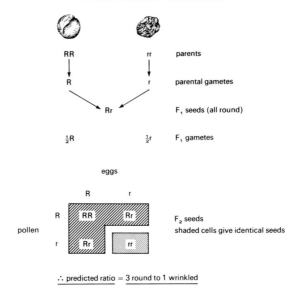

Figure 2.1   One of Mendel's experiments involving two variants of one trait.

to propose his second law, or the principle of independent assortment—
i.e. genes of different characters behave independently as they are assorted
into gametes. (The latter is far from universally valid but an appreciation
of this only came later.) One of the experiments Mendel carried out
on two-character inheritance is summarized, with his interpretation, in
Figure 2.2. The interesting result comes in the F_2 generation. There are
two possibilities: (1) that traits derived from one parent are transmitted
together, which would predict only two kinds of seeds in the F_2 generation,
round-yellow and wrinkled-green, in 3 : 1 ratio as before (Fig. 2.1); (2) that
traits are transmitted independently as illustrated in Figure 2.2. In the
latter case there should be four kinds of seeds in ratio 9 round-yellow
(dominant–dominant); 3 round-green (dominant–recessive); 3 wrinkled-
yellow (recessive–dominant); 1 wrinkled-green (recessive–recessive).
Mendel obtained 315 : 108 : 101 : 32 of the respective kinds in good agree-
ment with expectation (2).

2.1.2   Complications

As might have been expected, subsequent research showed the genetic
mechanism to be more complex than that envisaged by Mendel. Never-

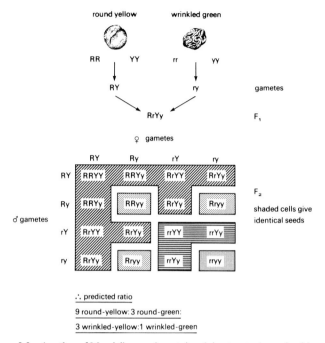

Figure 2.2 Another of Mendel's experiments involving two traits each with two variants.

theless, none of these complications alters the fundamental assumptions of Mendelism; i.e. that genes are non-blending and very stable. There follows a summary of some of the more important complications.

(1) *Dominance.* The feathers of the Andalusian fowl have a blue sheen, but amongst offspring there are always black and splashed white individuals. Actually the Andalusians are heterozygous for feather coloration, but black is not completely dominant over splashed white. This is not evidence for blending inheritance, because crosses between heterozygotes cause the speckled and black traits to segregate out according to expected Mendelian ratios.

We now know that the extent to which a gene dominates over another depends upon the way that it is expressed, not on any fundamental difference between genetic systems. In codominance both alleles are expressed equally (i.e. additively) and this is what happens in the Andalusian fowl. In partial or incomplete dominance one allele is inactive (not expressed), but the other expresses itself normally so that the phenotype gets half the dose of the effect. In complete dominance (what Mendel observed) one allele expresses itself as if it were equivalent to two, and in over-dominance one allele expresses itself to a greater extent than two of the same alleles in the homozygous condition. The picture is also complicated further, since the expression of alleles in one part (locus) of the gene complement may be controlled by the expression of alleles at other loci. This is known as *epistasis*. For example, if two homozygous mice are crossed, one with a dominant for an agouti (grey) coat and the other with a recessive black, all progeny should be agouti. Most are, but an occasional albino turns up, more frequently than can be explained by mutation. The reason is that a second pair of alleles at a second locus affects coat colour. The dominant allele at this locus is for normal coat as determined by the alleles at the primary locus. The recessive allele is for albinism, and any mouse homozygous for this will be albino irrespective of the state of the primary locus.

(2) *Linkage*. In Mendel's breeding experiments the factors investigated assorted independently, but we now know that genes are carried on chromosomes (see 2.2.1) and therefore that the assortment of each to the gametes depends upon their relative distribution between chromosomes. If two loci are close on the same chromosome they will be transmitted together.

    For example, Thomas Hunt Morgan, who initiated the genetic studies of fruit-flies (*Drosophila*) in 1909, found a deviation from Mendel's second law in two genes of *Drosophila melanogaster*; one affecting eye colour (pr = purple; pr^+ = red) and the other wing length (vg = vestigial, vg^+ = normal). (Note that the gene labelling convention is slightly different here—the mutant, usually recessive, allele being used to define the label and a superscript + being used to define the dominant allele. This was developed because large numbers of mutant alleles were discovered per locus in *Drosophila* and the convention is still used by *Drosophila* geneticists; other conventions have also been used and for a discussion see Ayala and Kiger (1980).) Morgan crossed heterozygous females ($prpr^+ vgvg^+$) with homozygous males ($prpr vgvg$). The expectation on the basis of classical Mendelian genetics is equal quantities of all four possible gene combinations: $prpr vgvg$, $prpr vgvg^+$, $prpr^+ vgvg$, $prpr^+ vgvg^+$ (Fig. 2.3a). However what he obtained was about ten times as many parental types ($prpr^+ vgvg^+$ and $prpr vgvg$) as the others. He explained this by suggesting that the two parental alleles were linked and hence transmitted on the same parental chromosome, as indicated in Figure 2.3b. But if the two genes are linked,

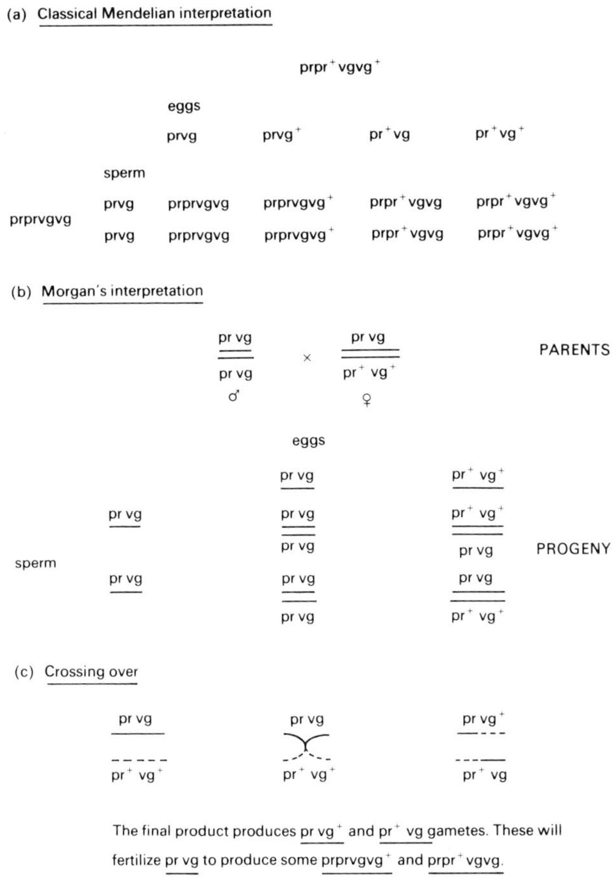

(a) Classical Mendelian interpretation

$prpr^+vgvg^+$

	eggs			
	$prvg$	$prvg^+$	pr^+vg	pr^+vg^+
sperm				
	$prvg$	$prprvgvg$	$prprvgvg^+$	$prpr^+vgvg$	$prpr^+vgvg^+$
$prprvgvg$				
	$prvg$	$prprvgvg$	$prprvgvg^+$	$prpr^+vgvg$	$prpr^+vgvg^+$

(b) Morgan's interpretation

$$\frac{pr\ vg}{pr\ vg}\ \delta \quad \times \quad \frac{pr\ vg}{pr^+\ vg^+}\ \female \quad\quad \text{PARENTS}$$

	eggs	
	$pr\ vg$	$pr^+\ vg^+$
$pr\ vg$	$\dfrac{pr\ vg}{pr\ vg}$	$\dfrac{pr^+\ vg^+}{pr\ vg}$ PROGENY
sperm		
$pr\ vg$	$\dfrac{pr\ vg}{pr\ vg}$	$\dfrac{pr\ vg}{pr^+\ vg^+}$

(c) Crossing over

$$\frac{pr\ vg}{pr^+\ vg^+} \quad\quad \frac{pr\ vg}{pr^+\ vg^+} \quad\quad \frac{pr\ vg^+}{pr^+\ vg}$$

The final product produces $pr\ vg^+$ and $pr^+\ vg$ gametes. These will
fertilize $pr\ vg$ to produce some $prprvgvg^+$ and $prpr^+vgvg$.

Figure 2.3 Interpretation of inheritance of eye colour and wing form in *Drosophila*. See text for further explanation.

how do gene combinations $prpr\ vgvg^+$ and $prpr^+\ vgvg$ ever form? The answer must be that genetic material is transferred between chromosomes (Fig. 2.3c) and this crossing over was later confirmed by cytological observations. Clearly, the nearer together two loci are on chromosomes, the more likely it is that they will be transmitted together and the less likely that they will be crossed over. Hence, measures of linkage begin to give information on the relative position of loci on chromosomes, and this is the basis of chromosome mapping.

In fact the first trait to be assigned to a chromosome location (by a series

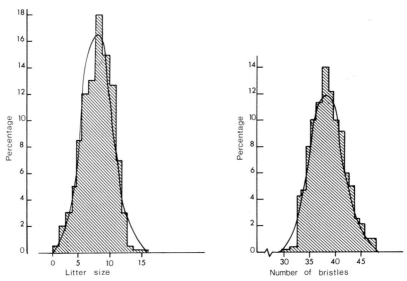

Figure 2.4 Frequency distribution of litter sizes in mice, and number of bristles on the ventral, abdominal surface of *Drosophila*. Both show approximate conformity to a normal (bell-shaped) curve. After Futuyma, D. J. (1979) *Evolutionary Biology*. Sinauer Associates, Massachusetts.

of key experiments carried out mainly in the USA between 1900 and 1910) was sex. In mammals and *Drosophila*, one chromosome, call it X, while present as two copies in the female is present as only one in the male. Instead of the other X chromosome, males contain a distinctive Y chromosome. Hence, females carry XX and produce eggs containing one X chromosome whereas males carry XY and produce sperm containing either X or Y chromosomes. If it is assumed that a gene or genes on these chromosomes controls sex, then it is possible to neatly explain the 1 : 1 sex ratio as follows:

		female XX		
		X	X	eggs
male XY sperm X		XX	XX	
Y		XY	XY	

(Note that this system is reversed in birds and some insects where the females are heterogametic and the males homogametic.) Sex ratios will be considered further in Section 3.7.

The X and Y *sex chromosomes* are morphologically distinct from the others (autosomes) and from each other. Hence, they are likely to contain unique alleles which cannot cross over from one to the other. This can lead to deviations from Mendel's predictions, and is known as sex linkage. For example, the allele for red eye colour in *Drosophila* discussed above is carried on the X chromosomes, as is an allele for white eye colour. The Y chromosome does not carry these alleles. Red is dominant to white. When Morgan crossed red-eyed females with white-eyed males, all the F_1 progeny were found to have red eyes. On crossing red-eyed F_1 males and females he obtained a $3:1$ ratio of red- to white-eyed flies, *but* all the white-eyed flies were male. The ratio of red-eyed females to red-eyed males was $2:1$. If we change the conventions from those on p. 14 such that X^W and X^w are the red and white alleles respectively, then it is easy to formulate an explanation of these observations:

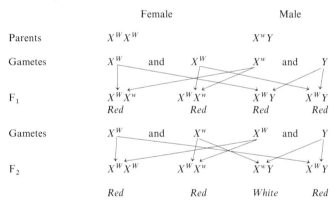

This association of genes with morphologically distinct chromosomes was very good support for the chromosomal organization of genes.

(3) *Continuously-varying characters.* Mendel concentrated on sharply distinct characters, but most are continuously-varying. For example, organisms do not usually fit into discrete size classes but vary continuously between the minimum and maximum sizes for the group. Moreover, the most frequent pattern of distribution is with the majority of individuals clustering to the mean for the character, and frequencies tailing off at the edges of the range (e.g. Fig. 2.4). For a time such variation was thought to

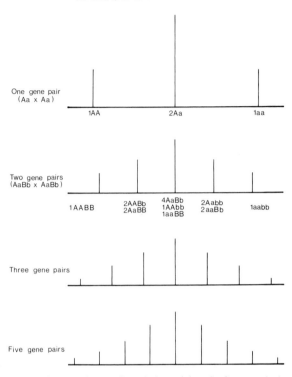

Figure 2.5   How continuous phenotypic variation might arise from particulate inheritance. Assume that the 'strength' of a character depends on the ratio of dominant to recessive alleles, then the greater the number of segregating loci specifying that character, the greater number of possible combinations of dominants to recessives and so the greater the number of possible phenotypic classes. After Strickberger, M. W. (1968) *Genetics*. Macmillan, New York.

fall outside the compass of Mendelian genetics, and W. F. Weldon (around 1900) even criticized Mendel's own results on this basis, for on closer examination most of the so-called discrete characters of the peas showed considerable variation; e.g. in the degree of wrinkling and greenness of the seed coat. In 1909, however, H. Nilssen-Ehle showed that quantitative characters could be controlled by Mendelian factors if a large number of genes were involved in determining a single trait (polygenes), each having a small additive effect—see (1) above. The rationale of this is illustrated in Figure 2.5. In general, the more genes that are involved, the more continuous is the variation.

   Over and above the genetic basis of variability in traits there is also an environmental effect. Size, for example, is crucially dependent on the conditions under which an animal or plant grows up. Probably all traits are a product of genetic and environmental effects so that normal distributions of the kind illustrated in Figure 2.4 are due to a combination of these factors. Only that component of this variability which is attributable to the genes is heritable, so this, expressed as a proportion of total variation, is defined as *heritability* (often symbolized as h^2). Measuring heritability is important for plant and animal breeders and can be discovered from the relationship between the characters of parents and offspring reared under the same conditions.

W. L. Johannsen, a plant geneticist, was one of the first to grasp (about 1900) the fact that the organism we see is a product of the environment in which it lives and the genes it contains. He therefore distinguished between the *phenotype* (the organism we see) and the *genotype* (the collection of genes it contains), and an appreciation of this distinction has remained important ever since.

(4) *Meiotic drive.* Alleles coding for the same trait are carried on separate chromosomes (homologous chromosomes) so each cell in a parent usually contains a number of pairs of homologous chromosomes (diploid conditions). These have to be separated in the formation of the gametes so that each gamete ends up with one member of each pair (haploid condition). This process of separation is known as *meiosis* and is the basis of Mendel's first law on segregation. It ensures that a heterozygote produces equal numbers of gametes containing each distinct allele. However, some gene-determined processes are able to distort this result. These can involve both pre- and post-meiotic events and are said to result in meiotic drive. This is reviewed by Zimmering *et al.* (1970).

One interesting example involves the so-called segregation distorter (SD) genes discovered in *Drosophila* (Crow, 1979). Here, while homologous chromosomes are paired, the SD genes influence their partners in such a way that they later form inviable sperm. SD genes of this kind become more evident when they are rare, because under these circumstances they are likely to occur as heterozygotes. As they become more common they pair with themselves more frequently and destroy themselves. Hence, SD genes are expected to occur at low equilibrium levels in populations.

2.2 Chromosomal and chemical basis of inheritance

2.2.1 The cytology of inheritance

The last three decades of the nineteenth century were of outstanding importance for research into the cellular basis of life. A series of workers, mainly German, discovered that the gametes were cellular and that the cell was not simply a homogeneous blob of matter, but consisted of a cytoplasm and a nucleus. With improved microscopes and special staining techniques they were able to look into the nucleus. There they found more heterogeneity: strand-like structures consisting of a substance that they called chromatin—the *chromosomes*. It was August Weismann (1834–1914) who first proposed that these chromosomes were good candidates for the material basis of heredity. He saw heredity as a transfer of material, what he called germ plasm, between parent and offspring, and suggested that a continuous and distinct germ track coursed through the ontogeny of the individual and 'spilled over' into the next generation leaving the old body soma behind. This he based on his observation that in many animals germ cells form and then separate off from other somatic tissues at a very early stage in embryology. Hence he distinguished between the germ line and the soma. In his *Evolution Theory* (1903; transl., Arnold, London, 1904) he further narrowed the germ plasm down to the nucleus of the germ cells, and then to the chromosomes and chromatin (first named by W. Waldeyer in 1888). This he did because histological evidence indicated that the chromosomes behaved just as they should if they were the heredity material—duplicating and dividing into two equal groups in the division of body cells; halving in number during the formation of gametes prior to the mixing of male and female contributions during fertilization. We now refer to these processes respectively as mitosis and meiosis.

The Mendelian factors, therefore, became located as *genes* (a term first coined by W. L. Johannsen in 1888) on the chromosomes. This kind of theory was of course quite distinct from the blending inheritance of Darwin. The germ line was separated from the soma and therefore from acquired characters. The genes were discrete and did not blend. Moreover, the chromosomal theory made more sense of some of the breeding experiments like those of Morgan on linkage in *Drosophila* (see 2.1.2).

2.2.2 *The chemistry of inheritance*

Genes may reside in or on chromosomes, but not all biologists were as confident as Weismann that they would turn out to be wholly chemical. T. H. Morgan himself wrote in his *Mechanism of Mendelian Heredity* (with A. H. Sturtevant, H. J. Muller and C. B. Bridges; Henry Holt & Co., 1915) that

> the supposition that particles of chromatin, indistinguishable from each other and indeed almost homogeneous under any known test, can by their material nature confer all the properties of life, surpasses the range of even the most convinced materialist.

Generating heterogeneity, or in other words genetic information, from homogeneity was the main problem—and the answer turned out to be that the chromatin was not as homogeneous as it appeared. Chemical probes showed it to be a nucleic acid consisting of repeated subunits (nucleotides) held together either by ribose sugar units (ribonucleic acids—RNA) or deoxyribose sugars (deoxyribonucleic acids—DNA).

The major breakthrough came in the mid-nineteen fifties when James D. Watson and Francis Crick deduced the structure of DNA from a combination of chemical data and X-ray 'pictures' of the molecule. Each nucleic acid consisted of two chains held together in ladder-like fashion by chemical bonds. The rungs of this ladder consisted of the nucleotides organized in such a way that only one (a small unit) could occur alongside a specific partner (a large unit). Hence the two sides of the ladder were said to be complementary. The ladder was twisted in the style of an electric flex to produce the now-famous double helix (see Fig. 2.6 for summary).

This deduction about structure immediately suggested a number of others about function. First, the sequence of bases on a ladder could occur in a variety of patterns and by these patterns could encode the genetic information. Chromatin was not homogeneous. Second, complementarity meant that if the sides of the molecular ladder were separated, each could act as a template of the other. Hence the information could be replicated and faithfully passed from cell to cell and from parent to offspring (Fig. 2.6d). Finally, the coiled structure meant that the vital, genetic information was protected from the vagaries of the outside world. The occasional accident might happen in replication, leading to what the classical geneticists thought of as mutation. Such events would be rare and might be caused by environmental factors (like irradiation) but would not be directed to any specific 'needs'.

This characterization that Watson and Crick worked out turned out to be substantially correct and led to a burst of activity in molecular genetics. The nucleus and nucleic acids were known to play a crucial part in protein synthesis, and because proteins are important and intimately involved in forming the fabric (as building blocks) and controlling the function (as enzymes) of the cell, it was easy to see how nucleic acids might specify the traits of the organism. Nucleic acids consist of replicated subunits of nucleotides, and proteins consist of amino acids, so the sequence of the former could code for the sequence of the latter. However, since there are more amino acids in proteins than nucleotides in nucleic acids (by approx. 5 : 1) the specification cannot be one to one. Theoretically, it is straightforward to show that only groups of three nucleotides (triplets) would

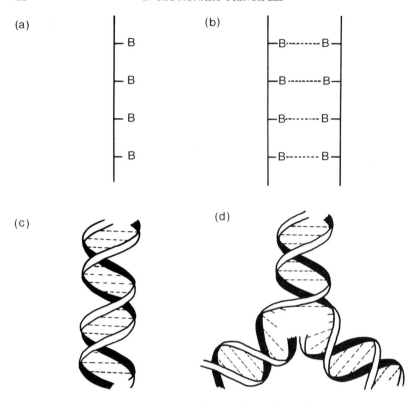

Figure 2.6   Structure of DNA. (*a*) Nucleotides (B) held together in chain; (*b*) complementary chains attached in ladder-like fashion by hydrogen bonds (broken lines); (*c*) ladder twisted in double helix; (*d*) replication—one chain unwinds to act as template in formation of two new chains.

provide sufficient variation (4^3 or 64 alternative combinations) for this purpose, and the triplet code was soon established as an experimental fact. The sequence of transfer of genetic information in protein synthesis in metazoa is two-stage from deoxyribonucleic acids (DNA) to ribonucleic acids (RNA) (known as transcription) and RNA to proteins (known as translation). A simplified summary of this is given in Figure 2.7.

2.2.3   *Relationship between classical and molecular genetics*

What part, or how much DNA is equivalent to a Mendelian gene? This is a natural question to ask but an impossible one to answer. First, it is to be

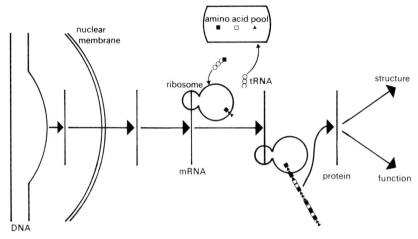

Figure 2.7 Transcription of DNA to messenger (m) RNA and the translation to proteins. The latter occurs on sub-cellular organelles, the ribosomes, and transfer (t) RNA molecules select specific amino acids from the cell pool and plug them into the appropriate places on the template. The protein may then enter into the fabric of the cell or act as an enzyme. After Calow, P. (1976) *Biological Machines*. Edward Arnold Publishers, London.

noted that nucleic acids do not code for noses, hair colours, etc. but for molecules which build and maintain such characters! Second, the way geneticists have carved up organisms into traits is arbitrary with respect to molecular constitution. For example, one molecule formed by one part of the DNA may occur in the nose, the hair and the eye. It was inevitable, therefore, that the gene should be redefined in terms of discrete parts of the genome, i.e. those which produce particular proteins. However not all genes code for proteins; some control the activity of other genes and so a distinction was made between structural genes (the former) and regulatory genes (the latter). Approximately 30 to 70% of mammalian DNA is repetitive, and if it has any function at all it is likely to be regulatory (but see Section 3.8). Finally, evidence is emerging that stretches of DNA coding for specific proteins may not be physically discrete and static. Genes and gene fragments can move around the DNA molecule and such transposable genes have been discovered both in microbes and metazoa. Hence the Mendelian concept of physically discrete (particulate) genes is becoming more fuzzy.

The link between DNA, Mendelian factors and organismic traits is not as straightforward as was once imagined but classical Mendelian observations can still be given a chemical basis. The earlier discussions on dominance and recessiveness are framed in such a way that the links with chemical reactions can easily be appreciated (see 2.1.2). Epistasis (2.1.2) can be readily reinterpreted in terms of interacting genes and gene products. There is no difficulty, for example, in seeing how gene expression might be influenced chemically both at the gene itself and also at the sites of cascading gene products by chemical factors in the internal and external environment. Finally, mutation can be interpreted as an error in the replication of the DNA. This may be confined to the replacement of one nucleotide by another (i.e. point mutation), or it may involve structural rearrangement of whole chromosomes (chromosomal mutation), and these may include deficiencies, duplications, translocations and inversions.

2.2.4 Lamarckism revisited?

Even before the structure of the nucleic acids was fully worked out it was realized that information should pass from the DNA to the proteins and not the reverse. This is known as the Central Dogma of Molecular Biology, and is in line with Weismann's idea that genetic information passes unidirectionally from germ line to soma. However, though the Central Dogma and Weismann's Principle are compatible they are nevertheless distinct. The former prohibits the flow of information from the cytoplasm of a cell to its nucleus and the latter from one type of cell (somatic) to another (germ line). Both prohibit Lamarckism. However, while the Central Dogma remains inviolate there seem to be ways, at least in principle, by which Weismannism might be circumvented. For example, viruses are known which can transfer nucleic acids from one bacterial cell to another in culture and, indeed, this is one of the basic techniques of genetic engineering. There is no reason why there should not be a similar cell-to-cell information transfer in many-celled organisms and hence why there should not be information transfer from somatic to germ cells. On this basis, E. J. Steele in his *Somatic Selection and Adaptive Evolution* (Croom Helm, 1980) has recently revisited Lamarckism.

Chance mutations are known to turn up in the somatic cells. Useful ones, i.e. which benefit the cell carrying them, may spread because they allow their bearers to divide more rapidly than other cells and this is particularly true of cells involved in combating disease. The more successful they are the more numerous they are likely to become. The mutant gene is therefore multiplied and because of this stands a better chance of being picked up by viruses and transferred to other cells and possibly to the germ line. There could be a period of somatic selection of somatic mutations in the body of an individual, therefore, prior to them passing through the Darwinian selection filter. Steele claims that this process would speed up evolution and render more understandable the evolution of complex and co-ordinated organs like the eye.

As yet, there is little experimental support for these ideas but they are at least plausible. However, there are also a number of difficulties with the theory. First, not all cells in organisms are capable of multiplying—for example this applies to the nerve cells and muscle cells—so somatic selection would not seem to be universally possible. Second, the spread of mutations through the somatic cells does not automatically mean that they are useful. Cancer mutants would do exactly that, but are lethal. Even if Steele's ideas do turn out to be correct, however, there is nothing mysterious about them and there are no grounds for re-invoking mysterious striving towards particular needs as a basis for explaining the diversity and adaptability of living things. Indeed it could be said that Steele's hypothesis is still based fundamentally on random variation and natural selection, the addition being that selection is allowed *within* as well as *between* organisms. As early as 1881, Wilhelm Roux was also suggesting in his book, *Der Kampf der Theile im Organismus* (Leipzig) that there would be an intra-organismic struggle (and hence selection) between tissues and organs for the limited resources supplied by feeding and that this would be an important force in evolution. Ironically, Weismann also accepted the possibility of intra-organismic selection, thought it might be important in developmental flexibility but dismissed it as an unimportant element of evolution.

2.3   Neo-Darwinism and population genetics

As well as solving problems for the Darwinian theory, Mendelism also raised a number of new ones. First, genetic properties such as dominance might, it was suggested, be more important in determining population characteristics than natural selection itself. After all, laboratory crosses between homozygous dominants and recessives lead to the numerical superiority of the dominants (see 2.1.1). Second, the discrete, particulate nature of heredity factors must mean that alterations in these factors are discrete and that each change is likely to have a large effect. Variation therefore occurs in big jumps rather than in the small, continuous way envisaged by Darwin.

De Vries, in his book *The Mutation Theory* (Open Court Publ. Co., 1909), distinguished between continuous, individual variation and discrete, saltational, variations. He applied the term mutation to the latter only (but as will have been apparent from what has been written before, the term is now used more generally for any spontaneous, random change, big or small). For De Vries it was these mutations that controlled evolutionary change and which were more important than natural selection itself.

The early Mendelians, like De Vries, therefore played down the efficacy of natural selection and emphasized the overriding importance of Mendelian ratios and mutations in moulding evolution. Some of the Darwinists were so impressed with this that they began to doubt the validity of Mendelism, and there ensued a bitter debate between these groups. Other evolutionists recognized the importance and challenge of Mendelism and set out to reconcile it with the Darwinian theory. Because the results of their efforts went beyond the mechanism that Darwin had proposed, the synthesis that emerged is often referred to as neo-Darwinism. Part of the solution involved shifting from considering the consequences of breeding from pairs of individuals to the consequences of breeding between individuals in populations; the Mendelian ratios thus became frequencies of occurrence of genotypes in the population and genetic and evolutionary changes were defined in terms of changes in these gene frequencies. Genetics therefore graduated to population genetics and this went hand-in-hand with the development of neo-Darwinism. This section reviews the main features of the programme and indicates how it attempted to answer the key questions raised by the Mendelians.

2.3.1 The Hardy–Weinberg result

In 1908, G. H. Hardy and W. Weinberg independently considered the following question raised by the Mendelians. Can gene frequencies change under the influence of the Mendelian constraints alone? To answer this they imagined a population with random mating (panmictic) and which was sufficiently large for statistical reliability. Notice that in this situation we do not know exactly what crosses will occur (unlike the Mendelian experiments) but we can guess what crosses are likely to occur on the basis of frequencies of individuals carrying particular genes. In this population, Hardy and Weinberg imagined two alleles, A and a, with initial frequencies p and q respectively, such that $p + q = 1$ (i.e. they account for all possibilities, there being no other alleles at this locus). The diploid genotypes are: AA, Aa, aa. Now they allowed random mating and so the *probability* of bringing two A gametes together to form AA is p^2. This follows from the basic theorem of probability theory that the probability of union of two consecutive events (say throwing two heads in tossing a coin) is equal to the product of their separate occurrence (i.e. $\frac{1}{2} \times \frac{1}{2}$ or $(\frac{1}{2})^2$ for the heads). By the same reasoning the probability of getting $aa = q^2$ and $Aa = pq$. The Mendelian constraint is that the heterozygote must be twice as frequent as each homozygote so that the expected frequencies are: $p^2 + 2pq + q^2$. The frequency of A in the new generation is $(p^2 + pq)/(p^2 + 2pq + q^2)$ which reduces to $(p(p+q))/(p+q)^2$ and, since $p+q = 1$, finally to p. But by definition the total frequency of genes at this locus is 1 so the frequency of the a alleles is $(1-p)$ which is q (again by definition). So the gene frequencies have remained exactly the same and will do so at every succeeding generation. The Mendelian mechanisms themselves, therefore, do not cause changes in gene frequencies and so they cannot be an important element in directing evolutionary change. Other factors must be involved.

2.3.2 Haldane, Fisher and the importance of selection

Can natural selection distinguish between small differences in genetically-determined traits which have only small effects on the survival and reproduction of their carriers? Could these small random changes cause significant alterations by themselves? Alternatively, must we assume that large saltational changes are the grist of evolution? These problems were addressed by J. B. S. Haldane in his *Genetical Theory of Evolution* (Longmans, 1930) and R. A. Fisher in his *The Causes of Evolution* (Oxford, 1932). These two books were immensely important in shaping the subsequent development of neo-Darwinism. Another important character in this was Sewall Wright. However, his approach was sufficiently different from that of Haldane and Fisher to merit separate treatment and we shall return to it in Section 2.3.3.

Both Fisher and Haldane expressed the intensity of selection in terms of changes in gene frequencies. Those alleles that increase in frequency *relative* to others are said to be fitter, so the change in these relative frequencies measures neo-Darwinian fitness. (Note that it is not sufficient to consider absolute changes, for an allele might decrease in absolute density but still leave more representatives in the population than other alleles if the latter decrease faster. It is then still fitter, since it is more successful than the other alleles under the prevailing conditions.)

The simplest starting point is the gene-dynamics of haploid populations (i.e. in which each individual carries one allele per locus) and Table 2.2 shows the basic calculations for two alleles (A and a) in such a population.

Table 2.2 Calculation of fitness in a haploid population

Allele	Number in one generation	Number at same time next generation	Net increase between time zero and t	Haldane's index of fitness (relative to A)
A	N_0	N_t	$\dfrac{N_t}{N_0} = R_A$	1
a	n_0	n_t	$\dfrac{n_t}{n_0} = R_a$	R_a/R_A

Haldane measured fitness by comparing the net rates of increase, R; i.e. by dividing throughout by the maximum R so that the fitness of the most successful allele becomes 1 and of the others becomes some fraction of 1. This index is usually denoted as w and is termed the *selective* or *adaptive value*. Sometimes it is represented as $1 - w$ ($= s$) which indicates the extent to which the gene is selected against, and is termed the *selective coefficient*.

This is an extremely simplified example. Generation times might differ between A and a and here it would be necessary to correct for this by dividing throughout by generation time. When generations overlap (such that the individuals in the populations at one time have different ages) and breeding is non-synchronous in the different generations, the situation is even more complex because there are then no standard intervals over which changes in the frequencies of A and a can be compared. Hence, Fisher preferred to measure fitness in terms of the relative rates of multiplication of genes in a population because in the extreme case of continuous breeding (some individuals of some generations producing offspring all the time) this can be deduced from the change in absolute numbers (say N for A and n for a) over any interval. Such multiplicative growth is modelled by $dN/dt = r_A N$ and $dn/dt = r_a n$ which integrate to $N_t = N_0 e^{r_A t}$ and $n_t = n_0 e^{r_a t}$ respectively (e = base of natural logarithms, r = constant, t = time and subscripts denote genes to which parameters refer). The coefficient r, what Fisher termed the Malthusian parameter (since Malthus had been one of the first to model multiplicative growth), is the measure of the rate of spread of each gene and therefore provides an index of fitness. Note that if we let t = average generation time as in the discontinuous case then

$$R_A = \frac{N_t}{N_0} = e^{r_A t_A}$$

and

$$R_a = \frac{n_t}{n_0} = e^{r_a t_a},$$

and if, for convenience, we let $t_A = t_a = 1$, $R_a/R_A = e^{r_a - r_A}$ and therefore $\log_e w_a = r_a - r_A$. This indicates that the two indices of fitness are related. Haldane's measure of fitness is usually preferred by population geneticists whereas Fisher's measure is often used by evolutionary ecologists. In the rest of this section we concentrate on w and s and return to r in Section 3.4. (For a more detailed discussion of r and its limitations as a measure of fitness, see Charlesworth (1980).)

The equations for the diploid case are more complicated than the haploid case, because the extent to which an allele spreads or recedes in a population depends upon which alleles it becomes associated with in heterozygotes. For example, a recessive deleterious gene will be protected from selection in heterozygous associations with advantageous dominants. Alternatively, selection against a deleterious dominant will lead to the elimination of advantageous recessives when they are associated with it in the heterozygotes.

It is possible to calculate the fitness of genotypes in much the same way as the constituent alleles. This is rather artificial, since homozygotes in one generation may contribute to the heterozygotes of the next and *vice versa* (i.e. genotypes contribute genes, not genotypes, to succeeding generations). Nevertheless, from these genotype frequencies it is possible to deduce the change in allele frequencies provided we have additional information on (*a*) the initial frequencies of alleles and (*b*) the extent of dominance. We do this using adaptive values and selection coefficients rather than Malthusian parameters but a similar procedure is possible with the latter (Ricklefs, 1980).

The simplest possible case involves a single locus with two alleles, say A and a, with respective frequencies p and q. Most of the theory in population genetics is based upon this type of system. Assume that selection shifts the Hardy–Weinberg equilibrium of a fictitious population as follows:

genotypes	AA	Aa	aa
frequency before selection	$0.16\ (p^2)$	$0.48\ (2pq)$ i.e. $p = 0.4, q = 0.6$	$0.36\ (q^2)$
frequency after selection	0.32	0.52	0.16

The change that has occurred for each genotype (i.e. equivalent to R in Table 2.2) is therefore $AA = 0.32/0.16 = 2$; $Aa = 0.52/0.48 = 1.08$; $aa = 0.16/0.36 = 0.44$, and the genotype fitnesses are then obtained by dividing throughout by 2 (i.e. the biggest value). Hence:

	AA	Aa	aa
W	1	0.54	0.22
S	0	0.46	0.78

(The upper case is used for W and S, to denote genotype rather than gene properties, and this convention will be used throughout.) Now working backwards, the new frequencies of genotypes after selection can be obtained from the product of initial frequencies and the Ws, adjusting each to a fraction of 1 by dividing by the sum of these individual terms. This sum is sometimes referred to as $\bar{W}$ and is the mean of the Ws.

	$W_{AA}p^2$	$W_{Aa}2pq$	$W_{aa}q^2$	
i.e.	1 (0.16)	0.54 (0.48)	0.22 (0.36)	
=	0.16	0.26	0.08	$\Sigma = 0.5$
new frequency	$= \dfrac{0.16}{0.5} = 0.32$	$\dfrac{0.26}{0.5} = 0.52$	$\dfrac{0.08}{0.5} = 0.16$	$\Sigma = 1.0$

Table 2.3 generalizes these results. From the initial frequencies of genes and the Ws we can now calculate the final frequencies of alleles (i.e. q' and p'). Hence the frequency of a after selection is:

$$q' = \frac{q^2 W_{aa} + pq W_{Aa}}{\bar{W}}$$

$$= \frac{q(q W_{aa} + p W_{Aa})}{\bar{W}}$$

The change in frequency (Δ_q) over one generation is therefore $q' - q$. These equations are usually expressed in terms of S rather than W but this is a simple matter of substitution. Similarly p' is simply obtained as $1 - q'$.

Table 2.3   Generalization of calculation of changes in genotype frequencies discussed in the text

Genotype	AA	Aa	aa	Total
Frequency before selection	p^2	$2pq$	q^2	1
Fitness	W_{AA}	W_{Aa}	W_{aa}	
Proportionate contribution	$W_{AA}p^2$	$W_{Aa}2pq$	$W_{aa}q^2$	$\bar{W}$
Frequency after one generation of selection	$\dfrac{W_{AA}p^2}{\bar{W}}$	$\dfrac{W_{Aa}2pq}{\bar{W}}$	$\dfrac{W_{aa}q^2}{\bar{W}}$	1

The rate and way that gene frequencies change from generation to generation depends on (1) the degree of dominance or recessiveness of an allele (which alters the relationships of the Ws—i.e. with dominance $W_{AA} = W_{Aa} \neq W_{aa}$, and with no dominance or over-dominance $W_{AA} \neq W_{Aa} \neq W_{aa}$); (2) which allele is favoured (e.g. as already mentioned a heterozygote will hide a deleterious recessive and slow down its removal, or expose an advantageous recessive to selective removal, thereby retarding its spread through the population); and (3) the intensity of selection. These criteria lead to different forms of the above equations describing changes in allele frequencies, and summaries can be found in most textbooks on population genetics. However, the solutions to these equations (what they mean for the rate and form of changes in frequency) cannot be deduced analytically (i.e. from 'looking at' the algebra) but only by specifying real numbers for substitution into them as q and W (i.e. by simulation).

The result of one such simulation, actually carried out by Haldane, is given in Table 2.4. This is for a case in which a dominant allele is advantageous and in which the selective pressure is relatively low. It indicates that the change in gene frequency is dependent on frequency. At very low frequencies the effects of selection are slower than at intermediate frequencies, and the rate slows down considerably as the allele approaches fixation (i.e. a complete take-over). This is because the less advantageous recessive is protected in the heterozygote and is difficult to remove completely. An advantageous recessive takes even longer to establish at low frequencies since it is removed from the population in the heterozygote by selection against the less advantageous dominant, but it approaches fixation more quickly. A number of simulations with $S \geqslant 0.001$ and various relationships between $W_{homozygote}$ and $W_{heterozygote}$ indicate that shifts in gene frequencies in the advantageous alleles from 0.1% to more than 50% of the population occur in less than 10^4 generations. We shall refer back to this figure later.

If the effect of selection is to increase the frequency of the favoured allele, then it must also be to increase the average fitness of the population as a whole (i.e. $\bar{W}$). Moreover, since the rate of change in frequency is slowest

Table 2.4 The number of generations required for a given change in frequency of a dominant allele, if dominants have 1000 offspring per 999 recessives; i.e. $W_{aa} = 0.999$ or $S_{aa} = 0.01$.

p	0.001 to 1%	1 to 50%	50 to 99%	99 to 99.999%
No. of generations	6920	4819	11 664	309 780

when one allele is at very low or high frequency then the change in $\bar{W}$ is likely to be greatest at intermediate frequencies—i.e. *when the mix of alleles is most even*. This is an intuitive statement of a principle which Fisher proved rigorously and thought to be a fundamental feature of evolution by natural selection. It is often referred to as *Fisher's Fundamental Theorem of Natural Selection*—i.e. 'the rate of increase of fitness of any *organism* at any time is equal to its genetic variance in fitness at that time'. Fisher formulated the expression badly because by 'organism' he clearly meant *population*. Genetic variance in W is greatest the more even the frequencies of alleles, and hence the spread of Ws.

Fisher referred to his theorem as fundamental because it precisely defined the relationship between two of the corner-stones of Darwin's theory—variation within populations and rates of evolutionary change. However, the relationship has turned out to be more ambiguous than Fisher suspected. For example, if fitness is itself dependent on genetic variance (as is plausible in certain situations; see 3.5), the rate of change of fitness need not be related in any simple way to that variance. Similarly, Fisher's result is derived for a one-locus two-allele situation, and when more than one locus is taken into account the relationship between the two variables can become even more complex (see 2.3.3). There are also other potential ambiguities and problems (Turner, 1970). Hence, the implication from Fisher's Fundamental Theorem that natural selection will always operate to *maximize* fitness cannot be accepted without reservation and this has implications for investigating adaptations (see Chapter 3). Another concept associated with the Fundamental Theorem is *genetic load*—the difference between the maximum potential fitness for a population and $\bar{W}$. This difference is introduced by any process which generates genetic variability including mutation (mutational load) and genetic reorganization by segregation (segregational load). As with the Fundamental Theorem, and possibly because of it, use of genetic load is not without dispute and controversy and will not be taken further here.

Turning now to the questions raised at the beginning of this section: how much more important is selection than *recurrent* mutation? Note that for an allele to increase in frequency by mutation and without selection the mutation must occur time and time again; i.e. it must recur. Consider a one-way mutation of A to a with a frequency of u. In general there will also be a back mutation of a to A, which will tend to balance the forward effect. However, let us ignore that and use the most favourable system for a frequency change by mutation. From observations on *Drosophila* and *Primula*, Haldane estimated a mutation rate of about 10^{-6} per gene and this estimate has been confirmed by subsequent work on other organisms. Haldane calculated the potential impact of mutation as follows: the frequency of A after one generation (p_1) is $(1-u)p_0$ (where $p_0 = $ initial frequency), and after two generations (p_2) is $(1-u)p_1$ or $(1-u)^2 p_0$ etc. Generalizing, $p_t = (1-u)^t p_0$, so by rearrangement the time t (in generations) to reduce p_0 to p_t is

$$t = \frac{\log_e (p_t/p_0)}{\log_e (1-u)}.$$

We can now use this equation to discover how many generations a mutation rate of 10^{-6} would take to shift the frequency of allele a (i.e. q) from 0.1 % to 50 %. If $p = 1 - q$, then the shift in A must be from 99.9 % to 50 %, so:

$$t = \frac{\log_e (50/99.9)}{\log_e (1-10^{-6})} \simeq 10^6 \text{ generations}$$

Remember that the equivalent figure for the action of weak selection was less than 10^4 generations. So even with the most favourable assumptions for mutations and with unfavourable assumptions for natural selection the latter proves to be a far more potent force of change (by almost two orders of magnitude) than the former.

This, of course, still leaves open the question of the importance of big as compared with small mutations. Fisher, in line with Darwin (see 1.5), argued that the probability that any mutation will have value is inversely proportional to the size of the mutation, since a large, violent change is unlikely to be compatible with the other traits with which the new one has to integrate. This is a question to which we shall return in Chapter 5.

2.3.3 Sewall Wright, adaptive landscapes and chance

The above analyses were based on a consideration of pairs of alleles situated at the same locus. However, most phenotypic traits are controlled by ensembles of genes involving many alleles and loci. These genes may be more or less strongly linked and subject to interaction (see 2.1.2). Hence the fitness of particular alleles at one locus is dependent on alleles at other loci and, indeed, will be influenced by the whole genetic environment in which they occur. Perhaps linkage groups, whole chromosomes or even genomes are more meaningful units of selection than individual genes (see 3.8). Unfortunately, the genetics of such polygenic systems is formidably complicated, so our grasp of these more realistic systems is weaker than that of the one-locus two-allele case. The assumption is that the basic principles will remain the same even if the details do not. However, there may be important exceptions.

Sewall Wright, another pioneer of neo-Darwinism, developed a conceptual framework for considering the interaction between genes and their influence on fitness (see Wright, 1931, *Genetics*, **16**, 92–159). This is referred to as an *adaptive landscape* or *fitness space*. In it each axis of an *n*-dimensional space defines the frequency of each gene in the population, and a final dimension defines the mean fitness $\bar{W}$ so that each combination of genes defines a particular fitness. Thus, in the simplest possible, three-dimensional model, two axes represent the frequency of two alleles at each of two loci and the space between these axes represents all possible

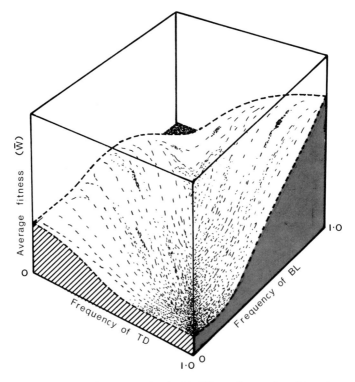

Figure 2.8 Adaptive landscape associated with different frequencies of two chromosome inversions in a grasshopper. BL = Blundell inversion; TD = Tidbinbilla inversion. After Shorrocks, B. (1978) *The Genesis of Diversity*. Hodder & Stoughton, London—data from Lewontin, R. C. and White, M. J. D. (1960) *Evolution*, **13**, 561–564.

combinations of these genes. The third dimension represents the $\bar{W}$ for each combination. Hence we have a three-dimensional landscape (the more complex situation generates an n-dimensional landscape which is as difficult to handle as it is to imagine). An example is given in Figure 2.8. In this model the fittest populations lie at the top of the 'hills' and the effect of selection is to cause populations to climb the hills and keep them there. Not all adaptive peaks will be or will have been occupied, since there are more potential combinations of alleles in a species than species themselves! Hence numerous peaks will never have been tried out and some will be higher than others. Which peak is climbed depends upon the proximity of populations and, therefore, on the initial genetic conditions before selection is applied. This is an important divergence from the one-allele two-locus case which predicts that $\bar{W}$ should always be maximized (i.e. only the highest peaks should be climbed), and the fact that this need not be the case has implications for questions of organismic adaptation that will be treated later (Chapter 3).

Figure 2.9 uses adaptive landscapes to illustrate, in a qualitative way, some evolutionary possibilities. (Note that we are observing the landscapes from above and therefore see contours.) A illustrates what would be expected from a relaxation of selection or an increase in mutation. The field occupied by a population spreads down the slopes of its adaptive peak and if this is sufficiently great the species may be impelled to another peak. B illustrates the effects of an increase in selection. This causes the population to withdraw to the highest level of its adaptive peak. C illustrates the effects of changes in the environment. These cause changes in the relief of the landscapes such that the adaptive peak may 'collapse' under the population leaving it in a 'valley' instead. Selection then forces the population up a nearby peak.

A and B underline the fact that selection is *continuously* required to maintain a population on a peak and this is referred to as *stabilizing selection*, i.e. stable morphologies and physiologies are maintained by continuous selection against mutants. Process C, though based upon the same mechanisms, has a different effect. It directs the population to a new adaptive combination of traits and is referred to as *directional selection*.

D represents a small population. Here *chance* happenings might cause a shift away from the adaptive peak *despite* selection and this is a possibility that Sewall Wright made explicit. In principle any individual may by chance meet with an accident or fail to meet with a mate, and so fail to make a contribution to the next generation, irrespective of how well it is adapted. No matter how successful they are potentially, the genes it carries

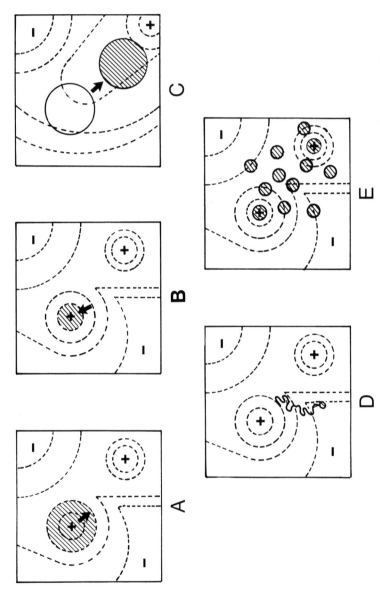

Figure 2.9 Plan views of several adaptive landscapes. A = relaxation of selection or increased mutation; B = stabilizing selection; C = directional selection; D = random drift; E = population divided into demes. After Waddington, C. H. (1939) *An Introduction to Modern Genetics*. George Allen & Unwin Ltd., London.

will therefore not be represented in the next generation. This will be unimportant in a large population. However, in small populations such chance happenings can have important effects on gene frequencies, since these are determined by small numbers of individuals. Under these circumstances the population may drift away from the adaptive peak and the process is referred to as *genetic drift*. Similarly, drastic reductions in population size or, what amounts to the same thing, the establishment of a new population by a few founders, may non-randomly sample the pool of genes constituting the original population. Hence an initial change will occur which is not related to selection. This is known as the *founder effect*.

E illustrates a large population divided into a number of sub-populations each occupying slightly different parts of the adaptive landscape. This is probably of common occurrence. The extent to which each group diverges from the mean genetic composition of the whole population either by selection or drift will depend upon the extent to which there is gene flow (i.e. migration of individuals) between groups. To oppose selection the rate of input of a gene must be greater than its selective loss and, to oppose drift, greater than the random loss. When the sub-populations are more or less isolated they are referred to as *demes*. Deme structure is therefore potentially important in evolutionary processes as will be discussed below (Section 3.6).

Genetic drift is sometimes referred to as the Sewall Wright Effect and is imagined to operate against selection, as in *D* above. However, as Sewall Wright realized, this effect, by releasing populations from local peaks, might allow them to cross valleys in the adaptive landscape and bring them under the influence of selective forces associated with new peaks. Hence, genetic drift can be an aid as well as a hindrance to adaptation.

2.4 Conclusions and summary

In following the historical progress of the genetical basis of evolution we have witnessed a number of important shifts in emphasis.

1. The Mendelians concentrated on the consequence of crossing *pairs of individuals* with known genetic constitution. The basic assumption was that characters were controlled by discrete factors that did not blend.
2. For Haldane and Fisher, evolutionary change was considered in terms of the changes in frequencies of *individual* genes in *populations*.

3. Wright's emphasis, however, was on *populations of genes* ($\bar{W}$s rather than *w*s) and Dobzhansky coined the term 'gene pool' to accommodate this idea (see his *Genetics and the Origin of Species*, Columbia University Press, 1937).
4. Starting with Weismann there has been continual progress towards exposing the physical basis of heredity. The ultimate result led to a redefinition of genes, mutations etc. in terms of DNA and protein synthesis.

(1) constitutes *Mendelian genetics*, (2) and (3) constitute *population genetics*, (4) constitutes *molecular genetics*, (2), (3) and (4) constitute *neo-Darwinism*.

At this stage it is worthwhile reiterating the beliefs of the majority of the neo-Darwinists. (I put it this way because not all neo-Darwinists share exactly the same beliefs and the differences in emphasis have been important in fuelling subsequent controversies.) The main assumptions are:

1. Genetic changes underlie evolution.
2. Mutations (in the modern, general sense) are the ultimate source of variation. These may be small (point mutations) or large (chromosomal) but the former are more likely to lead to advantageous changes than the latter and hence have been more important in evolution. The main point, though, is that they are random in the sense of not being directed.
3. The Central Dogma and Weismann's Principle apply.
4. Evolution is defined in terms of changes in gene frequencies.
5. These changes may occur by mutation, movement of genes into and out of populations, random drift and natural selection, but the latter is by far the most important cause.
6. These principles are sufficient to explain the diversity and adaptation of organisms on earth.

There are still a number of loose ends to which we shall return in later sections. Firstly, what of the relative contributions of genotype *and environment* in moulding the phenotype? (See Section 3.2 and Chapter 4.) Secondly, short-term changes in gene frequencies within populations say little about the long-term persistence of the populations themselves. Yet there is an intuitive feeling that the latter is at least as important a measure of evolutionary success as the former. Can this long-term persistence be defined in terms of neo-Darwinism? Is it even compatible with neo-Darwinism? Sections 3.6 and 5.4 will consider these questions.

2.5 Further reading

Good summaries of modern genetics are those of Gale (1980), Ayala and Kiger (1980) and Berry (1982). These are written almost entirely in terms of coefficients of selection, whereas, as already noted, Ricklefs (1980) gives a straightforward account in terms of the Malthusian parameter (see also Cook, 1971). More advanced texts dealing with populations with over-lapping generations are those of Roughgarden (1979) and Charlesworth (1980). The most straightforward account of quantitative genetics, and in particular of heritability (h^2), is still that of Falconer (1960; second edn., 1981). Molecular genetics is treated well in Watson (1976) and Woods (1980). A short, interesting history of genetics is given in Ludovici (1963), and an extremely personalized but nevertheless interesting account of the rise of molecular genetics in Watson (1980). The pace of molecular genetics is so rapid that it is difficult to keep in touch with all the developments. An unusual account which captures some of the excitement of this subject and describes a number of important developments is that of Watson and Tooze (1981).

CHAPTER THREE

ADAPTATION

3.1 Introduction

The last chapter defined fitness in terms of the relative spread of genes through gene pools, i.e. in terms of the *effect* of natural selection. The link between the spread of a gene and the gene itself, however, is dependent upon how it interacts with its environment, and this it does, not directly but through a gene-controlled intermediary, the phenotype. It is therefore relevant to consider why certain phenotypes are better at promoting gene-transmission (i.e. are fitter) than others under given ecological circumstances; i.e. the *cause* of selection. This approach is referred to as the adaptationist programme and combines a consideration of the form and function of phenotypes and their ecology and evolution.

Ever since Darwin, the adaptationist programme has been a popular pursuit for biologists, but there have been shifts in emphasis. As Dawkins (1976, 1982) has pointed out, for example, the Darwinists tried to interpret traits in terms of their good for the survival and reproduction of individuals (what Dawkins has called the *selfish organism* approach) whereas, since the neo-Darwinian revolution, the emphasis has been on explaining traits in terms of their good for the transmission of genetic units (i.e. genes and linked groups of them). This latter view leads to what Dawkins has called the *selfish gene* approach. Here phenotypic traits are to be evaluated strictly in terms of the way they influence the spread of the genes which control them. Finally, organismic traits have sometimes been interpreted in terms of the benefit they confer on the group (e.g. family, population, species) of which the organism is part. This might be referred to as the *unselfish organism* approach.

Most working adaptationists adopt a selfish organism approach which they assume is more or less compatible with the selfish gene philosophy—

interpreting phenotypic traits in terms of their impact on the survival and reproduction of their organismic carriers, which in turn ought to be positively correlated with the spread of genes coding for the phenotypic traits in question. This methodology is explained further and illustrated in Sections 3.2–3.5. The unselfish organism approach is discussed in Section 3.6. An area where both the selfish and unselfish organism approaches have been applied is discussed in Section 3.7 and, finally, we return to a consideration of the relationship between the selfish gene and the selfish organism in Section 3.8.

3.2 Methodology

There are two different but, as will become apparent, dependent ways in which the adaptationist programme has been pursued (Calow and Townsend, 1981).

Method 1 (*a-posteriori*/comparative approach). Here differences in characters are observed and *compared* in related species at different times in the history of the group (i.e. from the fossil record) or in extant fauna occupying different ecological circumstances. Then (*after the observations*) an adaptive explanation of the differences is formulated. With the time-correlated changes it is assumed that the changes in characters track alterations in ecological conditions and/or represent progressive adaptive improvements. For example, in the evolution of the horse (Fig. 3.1) the trends are explained in terms of a shift from life in forests to life on the plains. This involved (1) increases in the size and particularly the length and slenderness of the legs; (2) a decrease in the number of toes to one and an expansion of the 'toe-nail' of this remaining digit to form a hoof; (3) increase in the width of incisor teeth and the height of molars along with the development of complex patterns of hard enamel. (1) and (2) can be understood as adaptations for rapid locomotion over hard ground and (3) as an adaptation for feeding on coarse grass.

In the space-correlated changes it is assumed that populations of the same or related species have diverged in characters as a result of the different selection pressures they experience in the distinct ecological circumstances where they are found. It is therefore important that we should be able to precisely define the ecological differences which the organisms experience. Physiological ecologists, working with shore fauna, have large numbers of closely related species at their disposal, occupying different levels on the shore and subjected to differing degrees of exposure

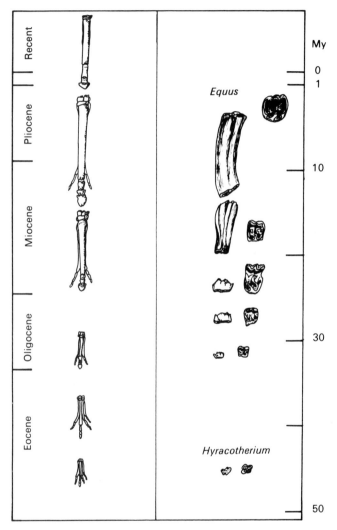

Figure 3.1   Trends in the evolution of the limbs and molar dentition of horses—see text.

to desiccation between tides and to wave action. Morphological, physio-logical and behavioural differences between species in different zones can often be explained as adaptations to these ecological differences and an example is given in Table 3.1. Here, the greater ability of the upper-shore *Patella vulgata* to survive high temperatures and desiccating conditions

Table 3.1 Some aspects of the physiological ecology of marine limpets (compiled from Davies, 1966, 1969)

	Patella vulgata	*Patella aspera*
Position on shore	upper levels	low levels
Survival at temperatures above 30°C	good	poor
Rate of water loss from tissues	low	high
Ability to withstand water loss	good	poor
Shell shape	▲	◣

than the lower-shore *P. aspera* can be considered as an adaptation to more frequent and longer exposure to the atmosphere between the tides. Moreover, the lower rate of water loss of *P. vulgata* as compared with *P. aspera* in air can be attributed to the differences in shell shapes between species (i.e. that of *P. vulgata* presenting less exposed surface to the desiccating atmosphere) and the morphological differences can therefore be understood in adaptationist terms.

There are a number of problems associated with Method 1.

(1) It is not always established if the differences in character are genotypic or just phenotypic. For example, the differences in shell morphology recorded in Table 3.1 might be attributable to the direct effect of wave action; i.e. in this case if both species were grown under the same conditions no differences would be observed. Non-genetic differences of this kind (sometimes referred to as phenotypic plasticity) are of little interest in the current context. Note, however, that the phenotypic plasticity may be genetically determined and may be adaptive. Examples of studies involving more precise genetic analyses will be given below (Section 3.3).

(2) A correlation does not automatically prove a cause–effect relationship. There may be hidden variables which are more important. Again the differences in the shell form of the limpets might be a response to differential predation which is itself controlled by wave-action or exposure between tides. Here the differences between traits would have to be explained in terms such as the extent to which a particular form of shell offers protection against predation rather than against desiccation or wave-action.

(3) Not all differences need be adaptive. Some could have arisen by chance (drift, founder-effect). Others might have arisen not by direct selection but as 'side-effects' of the traits which have selective advantage;

e.g. by pleiotropy (one gene having several phenotypic effects, not all of which need be adaptive) or allometry (see below, Section 4.2). Ageing might be just such a trait. It cannot be said to be obviously adaptive since, by definition, it reduces the vitality and increases the vulnerability of organisms which suffer it. It could be adaptive in a group sense, e.g. by making room for new mutations associated with the juveniles that replace the old individuals and hence stimulating evolution and long-term survival of the group—but such a group selectionist view is difficult to sustain (see Section 3.6). Alternatively, several theories relate ageing to pleiotropic effects. The gist of these is as follows: in nature most organisms do not die of old age, but by accident, disease and predation. As a result of these extrinsic factors, the probability of survival often decreases with age and so there is an age beyond which organisms are normally dead and do not make any reproductive contribution to the population. If one gene has several effects, some advantageous, others disadvantageous, there will be selection to delay the expression of the latter until that part of the life cycle where organisms are normally dead. The manifestation of these deleterious effects (intrinsic sources of mortality) becomes apparent when extrinsic sources of mortality are excluded, as in the laboratory or in human societies. On this view, ageing is not a specific adaptation but a 'side-effect' of genes which have unrelated, advantageous effects earlier in the life cycle.
(4) Characters might be adaptive and due to selection but there might still be no selective basis for differences amongst them. In other words, different adaptations can evolve as solutions to the same problems; i.e. there are multiple adaptive peaks in the fitness landscape and which one is climbed depends on the starting composition of the population. For example, this might be the basis of the difference between the Indian rhinoceros, which has one horn, and the African rhinoceros, which has two. Horns are adaptations for fending off predators, and it seems unlikely that one is specifically adaptive for Indian and two for African conditions. Rather, beginning with slightly different populations, with somewhat different developmental systems, it seems that the same selective forces caused achievement of different adaptive peaks (Lewontin, 1978).

*Method 2 (a-priori/*predictive approach). This programme is not concerned with explaining correlations between phenotypic traits and ecological variations but with *predicting* what these correlations should be. In principle, it makes the *predictions* before making the observations. *A-priori*, evolutionary arguments of this kind are often based on the assumption that natural selection is an optimizing process; i.e. there is some sense in which it can be said to result in the evolution of the best possible traits. Engineers and economists, who are also concerned with selecting best solutions to particular technological or economic problems, have developed mathematical methods, e.g. optimality theory, for dealing with this kind of problem and these have also been used in biology (Rosen, 1967). The main requirements for the application of optimality theory to any problem are (1) that all possible solutions to the problem should be known, and (2) that to each solution it should be possible to assign numbers or complex mathematical functions which denote either its value (v) or its cost (c) relative to some predetermined requirement. What the mathematics of optimality theory does is to search amongst these terms to find the largest v or smallest c.

The essential requirement for the application of optimality theory is the measurement of profits and costs, and this depends in turn on a clear, unambiguous definition of what is required of the system. For their systems, engineers and economists invariably have predetermined specifications against which the performance of any design can be compared. According to neo-Darwinian theory (see previous section) the phenotypic specifications of biological systems should tend to those that maximize the spread of descendants (carrying the trait), genes and perhaps even gene-complexes, and so profits and costs can be measured relative to these requirements. If we were able to define traits precisely in terms of their effects on survival, generation time and reproductive output, it would be a relatively easy matter to choose those that maximize neo-Darwinian fitness. Unfortunately, for all except a few traits, it is usually impossible to compute this value directly, and indirect assumptions are made about these relationships. This approach can therefore be summarized as follows: (1) assume that selection maximizes neo-Darwinian fitness (core hypothesis); (2) translate (1) into a phenotypic measure of fitness (auxiliary hypothesis); (3) using appropriate mathematical techniques, find the character which maximizes (or minimizes the reduction of) 2; (4) compare this prediction with what is observed in nature or what is found in contrived, experimental circumstances. In this programme adaptationists are rarely attempting to refute the core hypothesis. Rather they assume that this is more or less correct and then attempt to sharpen their understanding of the evolution of the phenotype by critically evaluating the auxiliary hypotheses.

Cohn (1954, 1955) and later Milsum and Roberge (1973) were interested in the adaptation of the vertebrate blood system. Their question was: what size of vessels should be favoured by natural selection? Now it is not

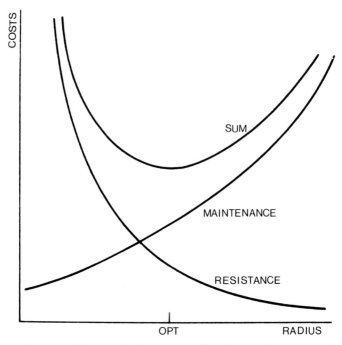

Figure 3.2   Cohn's model, predicting the optimal radius (OPT) for blood vessels. There are two cost functions; one increases and the other decreases with radius (see text). The optimum radius is the one which minimizes the sum of these costs—i.e. is the most economic solution.

possible to define vessel diameter in terms of its effects on survival, developmental rate and reproduction, so an auxiliary hypothesis is necessary, and the one chosen was that selection would be expected to favour the *most economic design*. The rationale of this is that the less resources used in building and maintaining the blood system, the more are available for combating disease and predators and for producing offspring. Cohn *et al.* envisaged two cost functions (Fig. 3.2), one which increases with reducing diameter of vessels (because frictional resistance to flow increases and necessitates more pumping power from the heart) and one which increases with increasing diameter (because more resources are required to build and maintain the larger system). The auxiliary hypothesis requires that the system with minimum costs is favoured, and this solution can be found graphically (Fig. 3.2) or from manipulating the equations specifying the cost-functions. Once numbers are inserted into the equations, specific optimum diameters can be calculated and compared with observation

Table 3.2   Cohn's predictions compared with actual observations

	Predicted	Observed
Radius aorta	0.43 cm	0.5 cm
Diameter of capillaries	2.2 μm	4.0 μm

(Table 3.2). Cohn *et al.* found reasonable concordance between their observed and expected diameters. Note, however, that in this exercise they were not trying to refute neo-Darwinism: rather, they were attempting to discover what it meant for the development of the blood system. The auxiliary hypothesis provides an ontogenetic rule that might be applicable to other systems; i.e. organ systems should be as economical as possible in form and function.

Method 2 also has problems. The main ones are concerned with defining the possible array of solutions from which the optimum might be chosen. On the one hand we cannot define these without reference to observation so Method 2 builds on Method 1, and this means that models might be moulded to fit rather than predict facts. On the other hand, not all conceivable solutions are feasible, so that some restriction has to be imposed on the imaginable solutions. That is, optimization models can be envisaged as searches through the adaptive landscape (Section 2.3.3) for highest peaks, but not all peaks will be available: some might define phenotypic organizations which cannot be built developmentally, and others might be excluded because of the initial genetic composition of the population before selection is applied. Alternatively, there might not be sufficient genetic variation in the population to allow the highest peak to be reached. Unfortunately it is extremely difficult to define these constraints *a priori* and the application of Method 2 might yield as much information on these as on adaptation itself.

3.3   Melanism in moths—*a posteriori* approach

With increasing industrialization in Britain, entomologists began to record the replacement of light colours in many different species of moths with dark or black ones. A particularly good example is the peppered moth, *Biston betularia*, where the dark form is referred to as *carbonaria*. The following facts were discovered: (1) the earliest records of dark forms came from sites near heavily industrialized areas; (2) the highest frequency of dark forms was found in sites near industrial centres (hence the phenomenon was described as *industrial melanism*); (3) melanics usually occurred in night-flying species. The *a-posteriori* evolutionary explanation is that light forms are more conspicuous to predators when they rest on sooty branches and tree trunks in industrial sites in the daytime. There are, of course, possible alternative explanations. In the first place, the melanism might be phenotypic; e.g. due to uptake of the industrial pollutants. However, breeding experiments established genotypic control and indicated that *carbonaria* usually segregates as if it were a dominant at a single locus. There are still other possible explanations; e.g. a dark colour might protect against the direct effects of pollution. To analyse the situation further, therefore, H. B. D. Kettlewell (1956, 1973) performed the following field experiments.

(1) To determine whether *carbonaria* had a greater fitness than typical peppered moths, Kettlewell released marked individuals into both industrialized and rural areas. More marked *carbonaria* were recovered in the industrial site and more typical forms in the rural area.

(2) Direct observations were made on predation rates. Similar numbers of *carbonaria* and typical moths were put on tree trunks in polluted and non-polluted areas. More melanics were eaten by birds in the rural areas and more non-melanics in the industrial area.

(3) One bird, the creeper, did not distinguish between the two moths. Kettlewell noticed that it feeds by creeping up and down the trunks of trees, and is therefore more apt to see the silhouette of the moth resting upon the bark of the tree than the colour of the wings.

(2) and (3) are particularly powerful supports for the idea that melanism has arisen in response to predation pressure.

This analysis on industrial melanism is based on Method 1 above, but was taken much further than is usual, or usually possible, by careful observation and experimentation. Moreover, it was possible to define the genetic basis of the change. Programmes such as these, based on Method 1, but taking into account ecological and genetic observations, are often referred to as *ecogenetics*—an approach pioneered by E. B. Ford (see Ford, 1975) and Dobzhansky (1937).

3.4 Examples of *a-priori* approach

3.4.1 *Application of optimality theory to life-cycle adaptations*

Most of the adaptations that have been considered thus far have either been morphological or physiological. Yet traits like developmental patterns, growth rates, size, life-span and investment in reproduction are also likely to have been subjected to natural selection. These traits are all concerned with bringing the products of reproduction into a condition whereby they can reproduce themselves, and they are referred to as life-cycle or life-history traits. Since the pioneering work of Lamont Cole (1954) much has been written about life-cycle evolution (for good reviews see Stearns, 1976, 1977).

Here, as an illustration, we concentrate on the amount of resources invested by a parent in making its gametes, and the post-reproductive life-span of the parent. The fact that not all organisms are explosive reproducers implies that there are important costs associated with reproduction. Otherwise neo-Darwinian fitness would always be maximized by maximum reproductive output and, therefore, by the maximum investment of resources in the production of the gametes. It is usually assumed that the more resources a parent invests in reproduction, the less are available for maintaining and protecting itself against accidents, diseases and predation, so the costs are in terms of parental survivorship. In consequence, as the investment in reproduction is increased, the post-reproductive survival chances of the parent decrease, and we here use Fisher's measure of fitness, i.e. r (see 2.3.2) in conjunction with a simple version of optimality theory to investigate what sort of compromise should be expected between these variables.

The fitness value, r, can be redefined as:

$$1 = \sum_{t}^{\infty} e^{-rt} l_t n_t \qquad (3.1)$$

where l_t = survivorship from birth until time t, and n_t = number of offspring produced at time t (for an easy derivation see Wilson and Bossert, 1971). In the simple case of annual breeding ($t = 1$), age-dependent fecundity (n) and post-reproductive survival, equation 3.1 becomes:

$$1 = e^{-r} l_a + e^{-r} l_j n \qquad (3.2)$$

where l_a and l_j are the survivorships of adults and offspring respectively, and n = number of offspring per parent. Hence:

$$e^r = l_a + l_j n \quad \text{and} \quad r = \log_e (l_a + l_j n) \qquad (3.3)$$

So from equation 3.3 we see that on a plot of l_a against n, isoclines of equal r would appear as straight lines with a slope of $-l_j$ (Fig. 3.3). However, because of the trade-off between parental survivorship and reproduction, not all combinations of l_a and n are attainable. A number of distributions

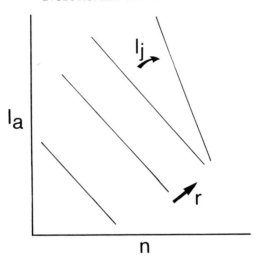

Figure 3.3 On plots of l_a against n, isoclines of r appear as straight lines with negative slopes of l_j. As offspring survivorship increases the slope increases. In principle, fitness, i.e. r, increases with l_a and n.

of attainable values can be imagined and a plausible one is shown in Figure 3.4. Here adult survivorship reduces more and more rapidly as more and more resources are invested in reproduction; i.e. more and more gametes are produced.

With these distributions it is easy to find the optimum solution, for it is that attainable combination of l_a and n which lies on the highest r-isocline; i.e. maximizes r or fitness. When, with the trade-off curve depicted in Figure 3.4, this optimum occurs at the extreme right of the curve, high reproductive output is favoured at the expense of the parent. This is known as a semelparous pattern—i.e. the parent breeds once and dies, like salmon. Alternatively, when there is a compromise between n and l_a, repeated breeding or iteroparity is favoured.

Now for a prediction. As the chances of offspring survivorship reduce (i.e. l_j decreases) the slope of the r-isoclines reduces and iteroparity is favoured. The reverse is true when l_j increases. (The exact outcome depends on the shape of the trade-off curve but we continue to assume that only that in Figure 3.4 is applicable.) To test this prediction we need to find either populations of the same species or of different, but closely related species, with different values of l_j. In fact there are surprisingly few precise data on age-specific survivorship for populations in natural situations, but there is some more or less anecdotal evidence which gives general support

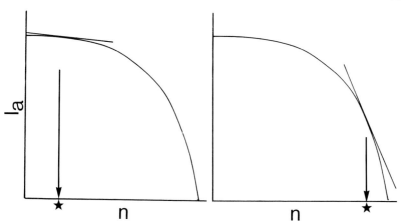

Figure 3.4 A possible form of the real relationship between l_a and n. In the case on the left, juvenile survivorship (l_j) is low and thus favours a small n (optimum marked with a star) and continued adult survivorship (high l_a). In the case on the right l_j is high, and a big n is favoured with a low l_a. The case on the left leads to iteroparity and that on the right to semelparity—see text.

to the predictions. For example, freshwater triclad planarians are born into poor trophic conditions in nature and it has been found (Calow and Woollhead, 1977) that the hatchlings of the semelparous species die less quickly than those of the iteroparous species when starved in the laboratory. Similarly, Blower (1969) has described semelparous and iteroparous species of millipedes in the British fauna, in which the former produce more offspring per parent than the latter. The semelparous species feed on leaves which are evenly distributed on forest floors, whereas the iteroparous species are more specialist and at least one feeds and lays eggs on patchily distributed logs. Dispersal of the young between resource patches is hazardous and so l_j here is lower than for the leaf-eating species. Iteroparity, with continuous or semi-continuous breeding, has presumably evolved for similar reasons in endoparasites but here, because the parent is surrounded by superabundant food resources in the form of the tissues of its host, it can produce vast quantities of progeny without having to pay the survival cost referred to above.

3.4.2 *Application of game theory to animal behaviour*

Another kind of constraint on what individuals do to maximize fitness is what other individuals do in the same population. Organisms, or at least the traits that they bear, can be thought of as playing games for existence with each other, and game theory, a branch of optimality theory, can help to define the optimum moves in these games. Using this approach, Maynard Smith (1972) has defined *evolutionary stable strategies* (ESSs) as those moves in the evolutionary game that cannot be beaten; i.e. traits or trait combinations that cannot be replaced by any invading mutant. ESS theory has proved extremely useful in the analysis of animal behaviour.

The 'behavioural game' is most vividly illustrated by aggressive inter-actions between players. Consider one such game that consists of two strategies (gene-determined sets of traits concerned with a particular behavioural pattern—i.e. moves in the behavioural game): *hawks* always fight to injure and kill their opponents even though they may risk injuring themselves, and *doves* display and never engage in fights. Which is the optimum move? This question has been considered in depth by Maynard Smith (1976). Assume that we can represent the effects of a particular move on fitness as a score and, for the sake of argument, let winners score $+50$ and losers 0; let the time wasted in display cost -10 and the cost of injury be -100. Now when a hawk encounters a hawk, half the time it is likely to win and half the time lose and sustain injury, so the overall pay-off is $0.5(50) + 0.5(-100)$, or -25. When a hawk encounters a dove the hawk always wins and gains $+50$; the dove loses and scores 0. When a dove encounters a dove, displays always occur, and it wins half the time and loses the other half. Hence the pay-off here is $0.5(50-10) + 0.5(-10) = +15$. We can summarize these calculations in a pay-off matrix (representing *average* pay-offs to an attacker):

Opponent	Hawk	Dove
Attacker		
Hawk	-25	$+50$
Dove	0	$+15$

What kind of behavioural system will evolve? In a population of all doves (average pay-offs $= +15$) any invading hawk mutants would do very well, getting an average of $+50$ at every encounter with a dove. Hence dove is not an ESS. Hawks would spread to take over the population, but then the average pay-off from each encounter would be -25 and a mutant dove would do very well, since though on each encounter with a hawk it gets a zero score, this is, nevertheless, better than -25. Hence, hawk is not an ESS.

Intuitively it would seem that a mixture of strategies ought to be stable. Let H be the proportion of hawks in the population and $1-H$ the proportion of doves. The average pay-off ($\bar{P}_H$) for a hawk is the pay-off for each type of encounter multiplied by the probability of meeting each contestant:

$$\bar{P}_H = -25H + 50(1 - H)$$

and for a dove:

$$\bar{P}_D = 0H + 15(1 - H)$$

These equations can be solved for H by making $\bar{P}_D = \bar{P}_H$ (which is the ESS) and this gives $H = 7/12$ and, therefore, $1 - H$ as $5/12$. This stable condition can be achieved by: (1) each individual in a population being hawk or dove (play pure strategies) and the population consisting of $7/12$ the former and $5/12$ the latter, or (2) each individual playing both hawk or dove (mixed strategy), the former with a probability of $7/12$ and the latter with a probability of $5/12$ at each encounter.

The hawks and doves game illustrates the application of ESS theory, using a simple and possibly unrealistic situation. Rarely, for example, are 'behavioural games' likely to consist of just two sharply distinct moves. Moreover, assigning scores to particular outcomes, though easy in principle is extremely difficult in practice and often involves the formulation of complex auxiliary hypotheses. Nevertheless, ESS theory has now been used widely in behavioural ecology and is giving considerable insight to complex 'behavioural games' (Krebs and Davies, 1981). We shall return to it in Section 3.7.

3.5   Variation in natural populations—implications for adaptation

Variation is introduced into a population by mutation (or possibly immigration). Once variation is introduced, the total array of genotypes can be considerably increased by recombination (and some see this as the main advantage of sexual as opposed to asexual reproduction—see Section 3.7), but this genetic variation is completely limited by the allelic variation available for recombination. Darwin saw natural selection as restricting this variation to an adaptive subset, and Fisher's Fundamental Theorem also implies evolution from greater to lesser variation. On this basis, populations are expected to evolve towards a homozygous *wild type* and to show little deviation from this adaptive optimum. Yet there is considerable variation, i.e. *genetic polymorphism*, in natural populations, and ecogeneticists have spent much time and effort documenting it.

The evidence for genetic polymorphism is of two main kinds.

(1) Direct observation of polymorphism in traits known to be gene-determined; e.g. melanism in moths, banding and coloration of the shells of snails, chromosome aberrations in *Drosophila*, blood groups in humans.

(2) Observations on enzyme structure using electrophoresis. Single amino-acid substitutions can alter the charge of the whole enzyme molecule sufficiently to alter its movement in an electric field. Hence similar molecules extracted from different individuals can be assessed in terms of this variability. The amino-acid substitutions can be explained in terms of nucleic substitutions and hence provide an index of genetic variability. The technique and rationale is illustrated further in Figure 3.5. The degree of polymorphism and hetero-

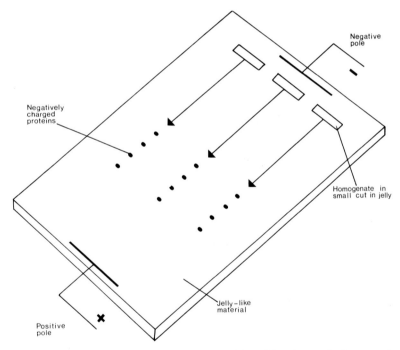

Figure 3.5 A horizontal electrophoresis gel. The gel is positioned on a plastic supporting plate, which is laid across the top of a tank containing appropriate buffer. Homogenates are put in the wells and an electric current is applied which causes the proteins in the wells to 'run' to an extent which depends upon their amino-acid composition (see text). Specific proteins can then be exposed using specific stains. The same protein from different animals may have slightly different amino-acid compositions and will 'run' to slightly different extents. After Shorrocks, B. (1978) *The Genesis of Diversity*. Hodder & Stoughton, London.

Table 3.3 Proportion of loci observed that showed enzyme variability and proportion of loci that were found in heterozygotic relationships. (Data compiled from Shorrocks, 1978.)

Species	No. of populations	No. of loci	Propn. of loci polymorphic	Heterozygosity per locus
Man	1	71	0.28	0.067
Mouse	4	41	0.29	0.091
Minnow	1	24	0.25	0.068
Starfish (Nearchaster aciculosus)	2	24	0.71	0.213
Horseshoe crab (Limulus polyphemus)	4	25	0.25	0.061
Drosophila pseudoobscura	10	24	0.43	0.128
D. willistoni	10	20	0.81	0.175

zygosity discovered from some electrophoretic studies is listed in Table 3.3.

Both techniques have suggested that there is considerable polymorphism and that this appears to be relatively stable. But classical Darwinism and neo-Darwinism predict the destruction of variability, and because of this there has arisen something of a tension between theoretical population geneticists and ecogeneticists. There are two possible solutions to the paradox.

(1) *Balance theory* argues that natural selection can stabilize variation if there is heterozygosity and the segregation of genes (Darwin could not have foreseen this from his blending theory). Balancing selection, as it is called, may occur through (*a*) heterozygote advantage—a superior heterozygote ($W_{AA} < W_{Aa} > W_{aa}$) will, by definition, keep homozygotes in existence; (*b*) frequency-dependent selection—here fitness is a function of the frequency of genes, such that as the frequency of one gene increases its fitness reduces and another gene is favoured until its frequency increases and the reverse occurs. This could arise from a predator always concentrating on common genotypes. Also mate selection could work like this; for example, *Drosophila* females, when offered a choice between mates, often appear to mate with the ones with rare characters; (*c*) changes in selection pressure in space (different genotypes being favoured in different parts of a habitat as in Fig. 3.6) and time (e.g. cyclical weather conditions favouring one genotype one season and another at another time).

(2) *Neoclassical theory* suggests that selection is either present, in which case it is directive, or it is not, in which case the alleles are selectively

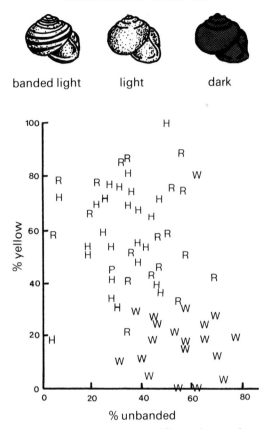

Figure 3.6 Changes in selection pressure between different places are known to have caused polymorphism in the shell of the land snail *Cepaea nemoralis*. This animal is eaten by thrushes which locate the snails visually. Banded light shells are difficult to see in low vegetation (herbage, R and hedgerows, H). In dark woodlands (W), uniformly dark and unbanded shells are the most difficult to see. After Luria, S. E., Gould, S. J. and Singer, S. (1981) *A View of Life*. The Benjamin/Cummings Publ. Co., Menlo Park, California. The original, classical work on banding in *Cepaea* was carried out by Cain and Sheppard (1954).

neutral. Much of the protein polymorphism, it is suggested, will be of this latter kind. It is known, for example, that protein molecules consist of a relatively small working part, whose structure is crucial for the way it works, and that the rest is 'packing'. Amino-acid substitutions in the latter will not affect function but will affect charge.

The controversy continues. Part of the problem is our inability to prove a negative—that is, being unable to demonstrate selection does not mean that it does not exist. Indeed, adaptive explanations have been found for protein variations where they were once thought not to occur. For instance, variation in the enzyme alcohol dehydrogenase in *Drosophila melanogaster* has been shown to influence the catalytic efficiency, heat stability, substrate specificity and specific activity of this enzyme. Hence the possibility of a balance caused by spatial or temporal variability in, for example, temperature and substrate availability becomes a real possibility. On the other hand, a few cases involving balancing selection do not prove its generality. Only a detailed investigation involving large numbers of loci and taking into account ecological as well as genetic factors will resolve this problem.

3.6 Adaptations for and of groups

It has been a fundamental assumption of the neo-Darwinian theory developed so far that the frequency of a gene increases if it specifies a trait which improves the survivorship and/or reproduction of the *possessor*. Yet there are many instances of animal behaviour that do not seem to contribute to the survival of individuals displaying the behaviour. For example, some castes of social insect are denied reproduction so that they can contribute more effectively to the work of the hive. A worker bee will kill an intruder, for the sake of hive protection, even if by so doing it kills itself. How can such altruistic behaviour evolve?

The solution came from a significant advance in our understanding of fitness developed in 1964 by W. D. Hamilton. The basis of the argument is that relatives carry similar genes, and the closer the relationship the greater the similarity. Therefore, if a trait is able to increase the survivorship of other individuals with the same trait, it may increase the frequency of the gene for that trait even if the donor is destroyed in the process. Altruism of this kind is likely to be greater between more closely related individuals (i.e. kin) and the process leading to it is termed *kin selection*.

Hamilton's contribution was to formalize this argument and to make it more precise (Hamilton, 1964). Suppose an altruistic trait is specified by a single pair of alleles, A and a, such that A makes an individual more likely to be altruistic — e.g. when an adult risks itself for the survival of juveniles by diverting the attention of a predator. In performing the act the adult reduces its survival chances by C and enhances those of the juveniles by B. Hamilton showed that A will increase in frequency relative to a only if the relationship (rl) between the donor and receiver of the benefit is close; more precisely, rl has to be greater than C/B.

Table 3.4 Coefficients of relationships, rl, between various relatives

	rl
Parent/offspring	0.5
Grandparent/grandchild	0.25
Full sibs	0.5
Half sibs	0.25
Uncle/nephew	0.125
Cousins	0.125

For example, in a parent-offspring situation the probability that any gene carried by a parent is carried by its offspring is 0.5, since the offspring carries the contributions of two parents. Hence $rl = 0.5$, and according to Hamilton's thesis, the altruistic trait must be capable of saving more than two offspring before A will increase relative to a. The rl-values in fact express the probability that organisms with a particular kind of relationship to each other share common genes, and hence the probability that the recipient shares the genes coding for the altruistic trait (some more examples are given in Table 3.4). As rl reduces, altruistic traits between the appropriate relatives must bring benefits from the actions of the donors to larger and larger numbers of recipients, otherwise the genes for altruism will not spread. Hence, and as already stated, co-operation is more probable between closely related individuals.

Hamilton's ideas even help to explain some aspects of the social organization of the Hymenoptera. Females in this group develop from fertilized eggs and so are diploid, but males develop from unfertilized eggs and so are haploid. In a population where both sexes are diploid, the coefficient of relationship between siblings is the same as between a parent and offspring ($rl = 0.5$, Table 3.4). However, as a result of the haplo/diploidy a female hymenopteran has more genes in common with a full sister than she has with her daughters. That is, sisters share all the genes that they receive from their fathers (he has just 1 set) and on average half the genes they receive from their mothers, so the coefficient of relationship between full sisters is $(0.5)(1) + (0.5)(0.5) = 0.75$, rather than 0.5. Hence, in this situation daughters should assist their mothers to provision and protect eggs rather than produce their own offspring, and so the hymenopteran genetic make-up predisposes them to evolve a social system in which sterile female workers care for their full siblings. There are even further implications, and for these the reader should consult Hamilton (1972). More recently, Hamilton's ideas have been applied to the study of social life in higher animals, birds, mammals and even man—see E. O. Wilson's *Sociobiology* (1975), and for a critical note, read Sahlins (1976).

Kin selection does not automatically lead to traits which are for the good of the species (i.e. for long term persistence, Section 2.4), only to those which are for the good of relatives. Nevertheless it is possible to imagine ways that such 'for-the-good-of-the-species traits' evolve by the selection of the groups themselves rather than the individuals within them. Imagine two semi-isolated populations of a species of predator. In one, a mutation arises to produce a 'super-predator' which is more efficient than the others at locating and capturing prey. This mutation must increase within the group even though it leads to the extinction of the prey and hence, ultimately, of itself. The mutation does not arise in the other population and so it persists and may ultimately take over the space occupied by the extinct group. A similar outcome would arise if a subgroup became isolated from the 'super-predators' but lacked the gene for 'super-predator'—e.g. because of the founder effect.

A slightly different example is as follows: imagine again the two sub-populations but this time containing immortal individuals. In one, a mutation for senescence emerges and this benefits the group because it limits the population, 'cleanses' it of worn-out individuals and increases the scope for favourable mutation. The group with senescence persists longer than the other. This example is different from the one involving super-predators, since it requires the evolution of a positive group trait, senescence, rather than the loss of the negative one of being a 'super-predator'. It is more difficult to imagine how this might work. How, for example, does the gene for senescence become established in the sub-population? It must establish itself either by recurrent mutation, drift or the founder effect since it cannot do so by selection—i.e. immortal parents leave more offspring than mortal ones, so the gene for senescence would be swamped.

What is the evidence for and against group selection?

(1) *Can it happen?* For group selection to be a viable possibility the rate of extinction of groups and/or the rate of formation of new groups (without the selfish gene or with an altruistic one) must be greater than the rate of influx of selfish genes by either mutation or, more importantly, gene flow between sub-groups. Some biologists are of the opinion that this requirement is too stringent for group selection to be of common occurrence (e.g. Lewontin, 1970) others are not (e.g. Gilpin, 1975, and D. S. Wilson, 1980).

(2) *Has it happened?* Is there evidence for traits which are essential for group persistence but which are unstable in individual selection? Senescence is one possibility and this is widespread. However it is also possible to explain this as a pleiotropic effect of advantageous genes which have been selected conventionally (Section 3.2). Another possible example is the ability of populations to regulate themselves at a level which does not over-exploit their resources. Wynne-Edwards, in his famous *Animal Dispersion in Relation to Social Behaviour* (Oliver & Boyd, 1962), provided abundant evidence of this kind of phenomenon. Yet it turns out, on closer scrutiny, that what appears to be self-regulation can often be explained in terms of density-dependent and independent regulation imposed from outside by predators, weather and even the resources themselves. Many other group traits have turned out to be explicable in terms of kin selection.

The tests for and against group selection are therefore inconclusive and because of this the controversy remains. That group selection can occur and has occurred in certain systems seems likely, but the question remains open as to the relative frequency of this as compared with individual selection. Group selection of a similar kind, but involving whole species rather than within-species groups, has also emerged as another possibility and will be discussed in Chapter 5.

3.7 Adaptations of and from sex

The essential features of sexual reproduction in most higher organisms are (1) the production of gametes by meiosis; (2) the recombination of elements in the genome by crossing over (see 2.1.2); (3) the random allocation of homologous chromosomes to each meiotic product (see 2.1.1); (4) the production of new individuals by syngamy (fusion of two gametes, usually from two separate individuals). Because of these processes, sex can generate considerable genomic variation between parents and their offspring even without mutation. This method of reproduction is widespread, occurring throughout the living world, and is probably very ancient.

There are also alternative methods of reproduction which involve single parents and gametes, and these are known collectively as parthenogenesis. There are several different types. In *arrhenotoky*, males are produced from unfertilized eggs, as in some Hymenoptera where males are haploid and females diploid. In *thelytoky*, females are produced from unfertilized eggs which are formed either by mitosis (meiosis being suppressed), a process known as apomixis, or by modified meiosis in which diploidy is restored after division by the fusion of the haploid nuclei (automixis). In apomixis the only source of variation is mutation, so offspring invariably resemble parents. In automixis, variation can be introduced by crossing over, but the post-meiotic fusion of nuclei increases homozygosity and, since this increases the chances that usually recessive deleterious genes are expressed, it is rare. Hence we concentrate on apomictic parthenogenesis, which is more widespread. It should also be noted that thelytoky can occur in association with sex in some species (e.g. rotifers and aphids cycle between sexual and parthenogenetic reproduction according to external conditions). Note also that the terminology used in this area is often confusing, since botanists and zoologists use different systems. That used above is zoological; for other terminology, see Maynard Smith (1978).

By far the most common and widespread mode of reproduction is the sexual one, and yet parthenogenesis seems to be the more efficient in terms of offspring production. Consider, for example, two females each with a similar physiology, but one with a mutation which leads to apomixis. Both produce the same number of eggs from the food they eat, but all the eggs of the parthenogenetic female are themselves female and parthenogenetic, whereas, on average, only half the eggs of the sexual female are female and the rest are male. In principle, therefore, the sexual system suffers a 50% disadvantage (the so-called cost of sex) compared with the parthenogenetic one in terms of offspring productivity, so the latter should spread through the population at the expense of sex. Hence sex is not an ESS (Section 3.4.2) and its widespread occurrence calls for some explanation.

There are a number of potential solutions to this puzzle and these have been considered by Williams (1975), Maynard Smith (1978) and Bell (1982). They can be divided broadly into those which rely on group selection and those that do not.

(1) *For-the-good-of-the-group explanations*. As already noted, sex generates diversity within populations. Moreover, through segregation it allows the dissociation of advantageous mutants from particular gene complexes so they can spread through the population, and also allows favourable gene complexes to be dissociated from disadvantageous mutations. Partheno-genesis, however, limits diversity (in apomixis it depends on mutation which is rare; see 2.3.2), impedes the spread of advantageous mutations (since mutations are locked into particular gene complexes and can only spread through individual clones) and means that disadvantageous muta-tions, because they are locked into genomes, will accumulate in populations

as they turn up in clone after clone. This latter process has been termed 'Muller's ratchet' after the man who first formalized the idea (Muller, 1954). Hence sexual populations should persist longer than partheno-genetic ones.

(2) *Neo-Darwinian explanations.* These focus on the effects of sex on the genes coding for it and the individuals which carry it. The offspring of sexual parents are likely to be more diverse than those of parthenogenetic ones. Hence, though parthenogenetic parents may be more prolific than sexual ones, their offspring are less likely to thrive and survive than sexual ones when the environment into which they are born is variable in space and/or time in terms of physico-chemical and/or biological conditions. Hence, these differences in survivorship may shift the balance in favour of sex.

Group selectionist explanations suffer from the same limitations as those already discussed in Section 3.6. Alternatively, the neo-Darwinian explanations have yet to be fully convincing that, in realistic environments, the genesis of variety is sufficient to pay the costs of sex. Unfortunately, the observational data are confused and confusing and no crucial tests have yet been formulated for decisively distinguishing between the many competing hypotheses (Bell, 1982). Hence the debate continues.

Whatever the outcome, however, it is beyond doubt that the evolution of sexual reproduction has had a profound and wide-ranging effect on the evolutionary processes themselves. A number of the more important effects are listed below.

(1) *Evolution of genetic systems.* The existence of sex has important implications for genetic mechanisms. Most fundamentally, it implies that genes (or linked groups of them) are more important units of selection than whole genomes, because the latter are ephemeral units, broken apart in meiosis and mixed together in syngamy. The mechanisms which bring this about, and others associated with them such as diploidy and domi-nance, have themselves been subject to selection and therefore invite adaptationist explanations. For example, diploidy and dominance might give some protection against the invasion of deleterious mutants and so the genes associated with controlling these traits would be likely to spread through populations. In this context, Fisher (1930) proposed that the degree of dominance evolved through the incorporation of modifier genes. Thus modifier allele M is advantageous if it causes allele A at another locus to be dominant over deleterious mutations (say A^1). A and M therefore interact epistatically (see 2.1.2). However there are difficulties

with this interpretation (reviewed by Sved and Mayo, 1970). For example, M would have only a very small advantage, since it would exert an effect only when coupled with AA^1 heterozygotes, which would be rare at first. Since M is likely to have other physiological effects than merely modifying dominance, evolution at this locus would be determined mostly by these other effects. However, as we shall see (Section 4.3), dominance need not have evolved specifically to suppress the effects of mutations, but may be a manifestation of a generalized capacity to resist changes to the developmental pattern whether caused by genetic or environmental disturbances. Issues such as these have been addressed ever since the 1930s, when Fisher's work and C. D. Darlington's *The Evolution of Genetic Systems* (Oliver & Boyd, 1939) were first published. Yet they have proved surprisingly intractable to rigorous treatment and experimental investigation. The links between adaptational cause and evolutionary effects are here so intimately entwined that they are sometimes difficult to recognize, let alone tease apart. This area is, nevertheless, an important and developing one and Bell (1982) has referred to it as *metagenetics*.

(2) *Evolution of gametes.* Primitively, all the products of meiosis in a population would probably have been similar in size (isogamous), as is the case in many unicellular organisms today. There would be some advantage in increasing the size of gametes, however, because this confers more resources on the developing zygote and hence increases its survival chances. At the same time, advantages are to be gained from small size because this means that more gametes can be formed for the same resources. If there is selection for the production of large gametes, there is immediately selection on small gametes to find and fuse with them because this way the genes of the small gamete gain advantages from the resources stored in the large one. There might also have been selection for the large gametes to resist fertilization by small ones so that they could associate with large gametes themselves to produce even larger zygotes. However, the survival return from zygote size is likely to follow the law of diminishing increments—increasing in ever-decreasing amounts as size is increased. Genes for large gametes could not therefore persist against invading genes for small size that cheat (i.e. by using the benefits supplied by large-sized partners without paying the costs). On the other hand, small-sized gametes could not persist against large-sized invading mutants because of the poor survival of small zygotes. This argument can be made rigorous (Parker *et al.*, 1972; Bell, 1982) and predicts that anisogamy, with small (male) and large (female) gametes, is the ESS. Anisogamy is, of

course, the rule in both the plant and animal kingdoms and is the basis of all other differences between the sexes.

(3) *Evolution of the sex ratio.* Having seen how 'maleness' (small-gamete producers) and 'femaleness' (large-gamete producers) might have originated, it can now be asked what ratios of the two sexes might be expected in populations? We know the answer, usually 1 : 1, but if the producers of small gametes can produce more per unit resource than the producers of large gametes, why is the sex ratio not loaded in favour of females?

One solution to this puzzle can be formulated in ESS terms and traces back to Fisher (1930). Suppose there are 100 females per male in a population, then every male has 100 times the expected reproductive success of the females. Hence a parent whose progeny are all sons would leave almost 100 times the grandchildren of a parent whose progeny were female-biased. Hence the female bias is not an ESS and exactly the same would be true of a male bias. Only when the sex ratio is 1 : 1 would the expected success of male and female be equal. Hence this is the ESS.

In fact this assumes that it costs approximately the same to produce sons and daughters. Assume, for example, that it costs more (say twice as much) to produce sons because they are bigger than daughters. When the sex ratio is 1 : 1, a son has the same number of children as a daughter, but because the sons are more costly they are a bad investment, since they reduce the total number of offspring that a parent can leave. It would therefore pay parents to invest in daughters. This causes a female bias, but as this increases, the expected reproductive success of the sons goes up. The balance is reached here when parents *invest resources equally* in the two sexes, not when they produce equal *numbers* of offspring. Hence when the cost of making males and females differs, a deviation from the 1 : 1 sex ratios should be expected. In the hymenopteran wasp *Polistes*, Metcalf (1980) found that a species with small females and large males (*P. metricus*) had a female-biased ratio, whereas in another, related species (*P. variatus*) sexes were equal in size and the ratios were not biased. Note, however, that the example refers to Hymenoptera where females can control the sexes of their offspring (see 3.6). In the majority of species, the sex ratio is determined by sex chromosomes and it seems likely that the sex ratio will be constrained to the 1 : 1 value by the mechanics of meiosis, even if there is sexual dimorphism in size. Hence here genetic constraints override the other forces of selection (but see T. H. Clutton-Brock, *Nature* **298**, 11–13, 1982).

(4) *Sexual selection.* Because males produce smaller, and hence more,

gametes than females and because of the normal 1 : 1 sex ratio, males often compete for females. The selection pressure for male ability to get a mate is high because the cost of failure is high. This kind of selection is known as *sexual selection* and was identified by Charles Darwin. It can work by favouring the ability of one sex (usually but not always males) to compete directly for the other and/or the ability of one sex to attract the other. Examples are given in Krebs and Davies (1981). They point out that the intensity of sexual selection depends on the degree of competition for mates and that this depends in turn on (*a*) the extent to which both sexes invest in offspring (i.e. the less one sex invests in offspring, the more competition there is between members of that sex for mates—this usually means males compete for females, but not always—e.g. sticklebacks) and (*b*) the ratio of males to the females available for mating at one time (as the latter reduces, e.g. due to asynchronous breeding, competition intensifies).

The aspect of sexual selection which most caught Darwin's eye, and which has remained in focus since, has been the evolution of excessively elaborate displays and adornments, like the peacock's tail. These may have some adverse effects on their bearer since they are costly to build and maintain and make them conspicuous, yet they have evolved frequently and sometimes into fantastic forms—why? As yet there is no fully accepted or acceptable explanation but one, first formulated again by Fisher (1930), offers a plausible possibility. He suggested that adornments might be selected just because they are attractive to females. If, in a polygamous population, the majority of females prefer the adornment, then a mutation for absence of the adornment cannot invade since mutants will not be accepted as mates, even though the sons they might produce have better survival prospects than adorned sons. The evolution of adornments is, therefore, self-reinforcing *once the adornment and the female preference for it have originated*. But the crucial question then is: How did adornments originate? There are many possible answers to this question. One is that originally the adornment was correlated with other traits and effects which did enhance offspring production and survival. Hence, females that chose mates with adornments ensured higher fitness for their own genes. Only as the adornment became accentuated by self-reinforcement did this correlation break down. This, of course, is an hypothesis which is difficult to test once self-reinforcing sexual selection has got under way.

3.8 Selfish DNA and genes in organisms

Throughout this chapter the distinction between the selfish organism and the selfish gene has been left vague, as it usually is in the adaptationist programme. Wherein lies the distinction? And how important is it? The two approaches can be contrasted starkly as follows:

(1) *Selfish gene theory* proposes that particular phenotypic (organismic) characters have evolved because they help genes to replicate themselves.

(2) *Selfish organism theory* proposes that the genes that become most common in a population do so because they help organisms to survive and reproduce.

Since genes 'use' organisms to replicate themselves, and since organisms are controlled by genes, these two points of view will usually be alternative ways of looking at the same thing. A gene which worked at the expense of other genes in an organism, and hence at the expense of the survival and reproduction of the organism as a whole, would not spread very effectively. This implies that, within the context of the organism, there is likely to have been selection for integration and co-operation between genes. On the other hand, selfish gene theory does predict that genes that are neutral with respect to the survival and fecundity of organisms can spread. For example, the large amount of non-functional DNA referred to in 2.2.3 might be explained in exactly that way. Dawkins (1976) supposes that this surplus (sometimes called junk) DNA is like 'a parasite or . . . a harmless but useless passenger, hitching a ride in the survival machines created by other DNA'. The crucial question here, however, is how neutral particular genes and their effects are, and how neutral they need to be before they escape the scrutiny of selection within the context of the organism. For example, is the cost of replicating and carrying around the so-called junk DNA insufficient to affect the survival and fecundity of the carrier? Has it really no function, or is it that we have not discovered one yet? Some of these issues are debated in a series of lively articles published in *Nature* in 1980 (see **284**, 601–607; **285**, 617–620; **285**, 645–648).

Having become associated in organisms, it is easy to see why genes must co-operate, but another question raised by the selfish gene theory is: Why organisms? Why, asks Dawkins (1981) in the final chapter of his *Extended Phenotype*, have genes ganged up into genomes, and cells into multicellular bodies? The general answer is straightforward—genes in groups are transmitted more effectively than ones which are not—but the detailed reasons for this are not so straightforward and must now be brought under the scrutiny of the adaptationist programme. Moreover, if integration, co-operation and coadaptation are so important, why has this not resulted in more gene linkage? Why, to return to Section 3.7, have the mechanisms of sex, which are just as liable to break up favourable collections of genes as to create them, been allowed to persist? Why, in the words of Turner (1967), does the genome not congeal? These are difficult but, nevertheless, important and profound questions which the adaptationist programme is only just beginning to address.

3.9   Further reading

For a critical approach to the adaptationist programme see Gould and Lewontin (1979). For a justification of it, read Dawkins (1982). Use of optimality theory is illustrated further in Alexander (1982), and game theory in Maynard Smith (1982). Application of the adaptationist programme to physiological ecology is treated in Townsend and Calow (1981) and behavioural ecology in Krebs and Davies (1981).

CHAPTER FOUR

EVOLUTION AND DEVELOPMENT

4.1 Introduction

The neo-Darwinian approach to evolutionary biology concentrates on the genetic basis of population change and constancy, whereas adaptationists concentrate on phenotypes. However, these two aspects ought not to be separated, for the expression of the phenotype depends, in part, on the genotype, and the spread of genes depends upon how successfully phenotypes interact with their environments. The interrelationships are illustrated in Figure 4.1.

The main problem is that the relationships between genotypic and phenotypic effects are very complex and hence only poorly understood.

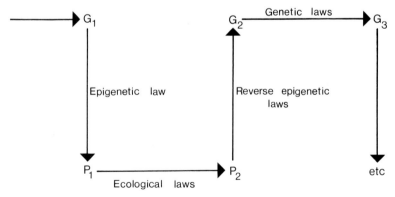

Figure 4.1 An initial distribution of genes (G_1) is transformed to phenotypes (P_1) according to epigenetic laws. The distribution of phenotypes changes according to ecological laws (to P_2). The phenotypes contribute genes to the gene pool (G_2) and these become distributed according to genetic laws (to G_3). This chapter focuses on the transformation of Gs to Ps.

The genes express themselves through a developmental system which is also influenced by environmental variables. Waddington (1975) coined the term *epigenetics* to describe the study of these interactions. We learnt something about the ecological and reverse epigenetic laws referred to in Figure 4.1 in Chapter 3. Here we turn to the epigenetic laws themselves. Just two of the most important principles will be considered: how small mutations might be amplified by development, and how the developmental system resists certain kinds of genetic change.

4.2 Developmental amplification

Joseph Needham (1933) distinguished three fundamental processes of ontogeny: (1) growth—increase in spatial dimensions and in weight; (2) development—differentiation and the increase in complexity of organization (morphogenesis); (3) maturation—the cessation of development and growth and the onset of reproductive processes. Julian Huxley (1932) investigated how (2) interacts with (1) by monitoring changes in the size of organs and relating these to each other and to changes in the size of the organism as a whole. Such relationships often conform to a simple mathematical rule:

$$Y = \beta X^\alpha$$

where Y = size of part; X = size of another part or the whole; α and β are constants (but *cf.* Smith, 1980).

$$\therefore \log Y = \log \beta + \alpha \log X$$

and

$$\log Y = b + \alpha \log X$$

where $b = \log \beta$. These are known as allometric relationships, and α is known as the constant of allometry: if $\alpha > 1$, Y increases in size more rapidly than X, and *vice versa* if $\alpha < 1$; if $\alpha = 1$, X and Y grow in direct proportion, and this special case is known as isometry. Examples are given in Figure 4.2. These relationships might reflect underlying developmental constraints and/or scaling effects and/or competition between parts of the organism for limited resources supplied in the food (Smith, 1980).

Needham (1933) pointed out that, both in principle and practice, the three fundamental processes can be altered in rate and timing with respect to each other, and that such adjustments have profound effects on the outcome of development. It is possible, for example, to accelerate differentiation with respect to growth experimentally and obtain dwarfs

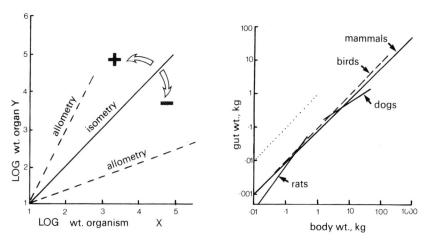

Figure 4.2 Examples of allometry in theory (left) and practice (right). Logarithmic plots of the size of one organ (Y) against another or of the organism as a whole (X) often give straight line relationships. When these have a slope of 1, the condition is referred to as isometry (for comparisons, denoted as a dotted line on the graph on the right) and X and Y increase in size at similar rates. The relationship between the gut weights and body weights of mammals and birds are roughly isometric. When the slope is greater than 1, the relationship is said to be a positive allometric one and Y increases in size more rapidly than X (e.g. data for rats). When the slope is less than 1, the relationship is a negative allometric one, and Y increases more slowly than X (e.g. data for dogs). These plots may be carried out for individuals within species (e.g. data for dogs and rats) or for individuals from several species (e.g. data for mammals and birds).

(= experimental dwarfism). For some invertebrates, it is also possible to inhibit reproduction and to obtain giants (parasites often castrate hosts and induce this same effect). If growth and development can be dissociated artificially in this way, then such *heterochrony* might play an important part in evolution. Small mutations might have small effects on the rates and timings of organ and tissue growth but large effects on the end product—in other words, small adjustments in α and/or b can have very important effects on the organization and morphology of the adult.

Alberch *et al.* (1979) have classified the possible ways that these adjustments might occur relative to a temporal component which specifies the initiation of development, age and the onset of maturation (i.e. process 3 of Needham's triad). Figure 4.3 summarizes this classification. Each square contains a *developmental trajectory*; i.e. an index of shape against size or age. (If size is used, then these figures can be thought of as equivalent to the relationships considered by Huxley.) The solid lines show the ancestral trajectory and the broken line the descendant. *Start*

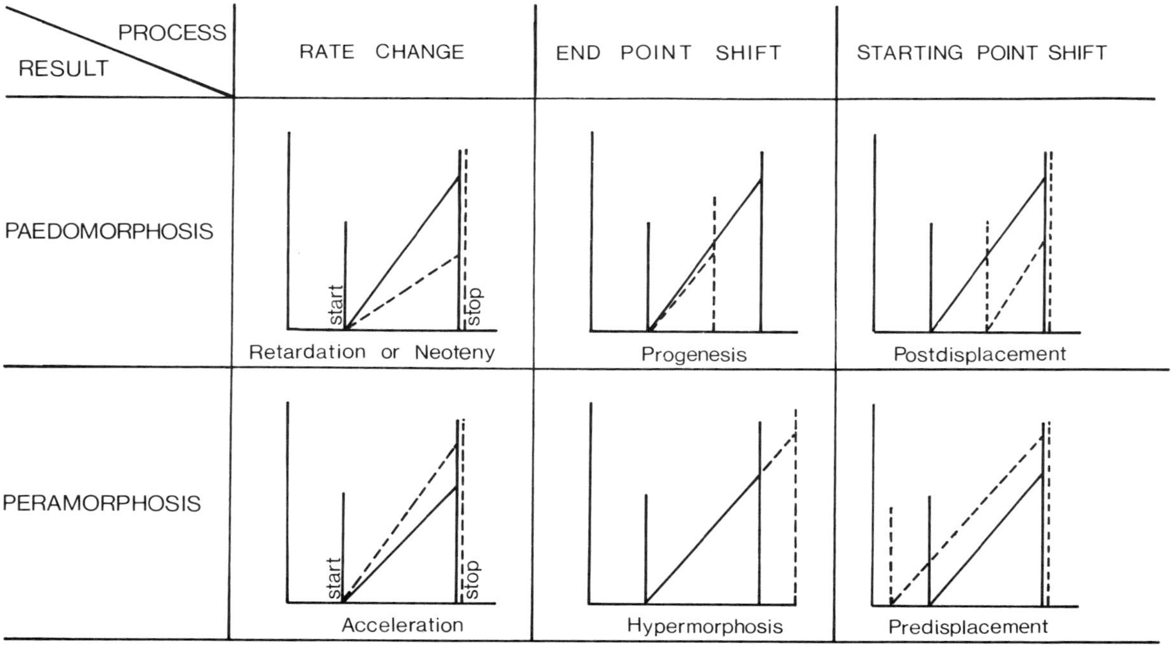

Figure 4.3 Classification of possible evolutionary shifts in development. Abscissas represent developmental time or size; ordinates represent morphological changes. See text.

represents the initiation of development and *Stop* the cessation of development. In all alterations in the top line, development is decelerated or truncated relative to the ancestor so that maturity occurs at an earlier developmental stage. This process is called paedomorphosis. In all alterations in the bottom line the reverse is the case and these are called peramorphosis. Here descendants go through ancestral stages at an earlier stage in development; i.e. the ancestral forms are recapitulated.

Examples of each of the component processes are as follows.

(1) *Retardation/neoteny*. The classic examples are certain amphibians (salamanders) in which gills and other larval organs are retained by the adult. The adult therefore remains tadpole-like and is aquatic.

(2) *Progenesis*. In some male crustaceans, sexual maturity is attained at a small size. These miniature males attach 'parasitically' to the much larger (sometimes by orders of magnitude) females.

(3) *Predisplacement*. In the 'creeper fowl' mutant the limb bones differentiate later than normal but then grow at the same rate. The time of hatching does not vary, so the emerging chickens have short legs.

(4) *Acceleration*. In certain ammonites the growth of sutures on the shells of descendant species is accelerated with respect to ancestral ones. However, this also involves other processes and will be dealt with more fully below.

(5) *Hypermorphosis*. In the now extinct Irish elk, the massive size, relative to other species, was probably achieved by maintaining growth and deferring reproduction.

(6) *Postdisplacement*. This and predisplacement might be important in controlling striping patterns in zebras. Bard (1977) postulates a single striping mechanism in the embryo for generating the vertical stripes. Depending on the stage of growth at which this mechanism is initiated, head stripes will be wider (process starts earlier = predisplacement) or narrower (process starts later = postdisplacement) and the patterns of striping can be varied considerably. Few patterns are, in fact, observed and these can be explained by the initiation of the stripe-producing process during the third (*Equus burchelli*), fourth (*E. zebra*) or fifth (*E. grevyi*) weeks of development.

However, we should not always expect 'pure' changes of the kinds described above and, indeed, morphological alterations are likely to be produced by a combination of developmental events. The allometry of suture length and body size in the ammonites already alluded to above and illustrated in Figure 4.4 provides a good example. Here the descendants' slopes are steeper than the ancestor's (Fig. 4.4, no. 1) so the differentiation

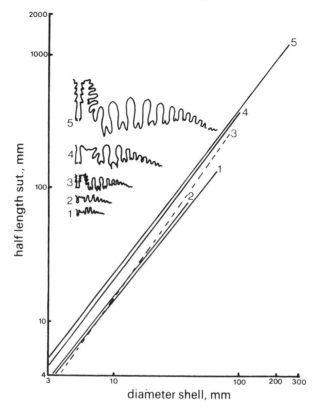

Figure 4.4 A mixture of developmental changes in the evolution of ammonites. The sutures (SUT.) of the ancestral (no. 1) and descendant (nos. 2–5) species are illustrated. The older part of the suture is the small tail to the right. See text for further discussion. After Newell, N. D. (1949) *Evolution*, **3**, 103–124.

of these sutures is *accelerated* relative to the ancestor's. Also, though, there is marked phyletic size increase in the group, and descendants continue sutural allometry well beyond the size of the ancestral adults (*hypermorphosis*). Finally, the descendants' trajectories are higher than those of the ancestor so the descendants begin with a head start, probably through some *predisplacement*.

To summarize: all the above examples show how modifications in developmental processes can have significant effects on the final form of the adult. In principle this can be effected by small changes in the rates of a few key processes and hence by small genetic changes. This is what is meant by developmental amplification.

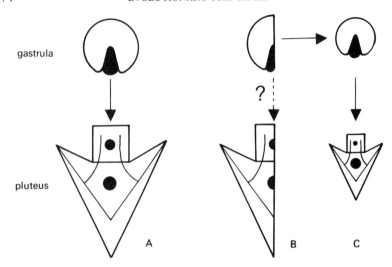

Figure 4.5 Experiments by Driesch on sea urchin embryos. *A*—normal development; *B*—Driesch cut the gastrula in half and expected a 'monster' to develop; *C*—instead the embryo reorganizes itself to produce a smaller, but nevertheless complete pluteus which can go on to form a normal adult.

4.3 Developmental canalization

One of the overwhelming conclusions that emerges from a study of development is that it is an ordered process which shows considerable powers of resisting disturbances imposed from both the 'outside' (e.g. by experimental manipulation) or from 'within' (e.g. by mutation). The following examples illustrate this point.

(1) Experimental embryologists have recognized for some time that the developmental processes of certain organisms are capable of producing well-formed, adaptive phenotypes even in the face of considerable out-side disturbance—a process known as developmental homeostasis. For example, Hans Driesch, a nineteenth-century biologist, bisected sea-urchin embryos at gastrula and pre-gastrula stages and observed that the halves were still capable of producing smaller but perfectly proportioned adults (Fig. 4.5).

(2) Most mutations are recessive. Morgan in the 1920s noted that all individuals in a given species of *Drosophila* in the wild looked remarkably alike, and interpreted this to mean that all individuals had the same *wild-type* genotype—this after all was what Darwinism predicted (see 3.5).

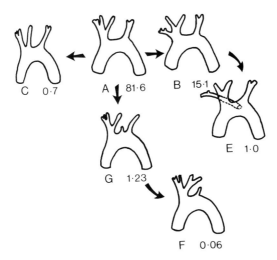

Figure 4.6 Variations in aortic arches of rabbits discovered in 3000 postmortems. There are six major types which account for 99.7% of the overall variation. Numbers are percentages and represent frequencies of encounter. After Sawin, P. B. and Edmonds, H. W. (1949) *Anatomical Record*, **96**, 183–200.

Dobzhansky (1937), however, showed this to be incorrect; identical-looking parents can contain different genotypes, since part of the genetic variety is masked by dominance. And, indeed, dominance itself may have evolved for exactly that reason; i.e. as a manifestation of a mechanism for dampening down the impact of mutations (see 3.7).

(3) Edmonds and Sawin (1936) and Sawin and Edmonds (1949) studied anatomical variations in the aortic arch of the domestic rabbit. Out of 3000 post-mortem examinations, they observed only 20 types of variations which can be reduced to 6 basic categories (Fig. 4.6). These did not occur with equal frequency. Such observations suggest a limited and non-random variation in complex organ-systems. Edmonds and Sawin found that the different types of aortic arch are the products of differential growth rates between the anterior thoracic region and the axial skeleton.

(4) Directional selection on particular traits in the laboratory often shows the following pattern: after an initial rapid response a certain threshold is reached and no further change can be achieved despite strong selection pressure (Fig. 4.7). Such resistance by the phenotype to strong external pressure by selection might occur because the variation in the selected population is quickly used up and/or because of the involvement of

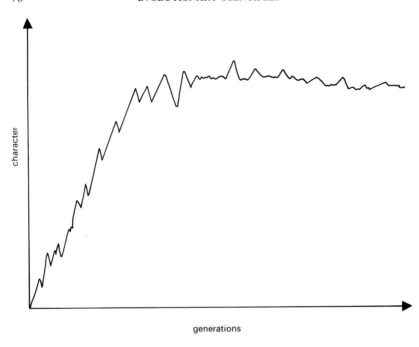

Figure 4.7   A common pattern for the effects of directional selection—see text.

internal, genetic and developmental regulation. The latter is thought to be important.

(5) The proboscis of *Drosophila* will sometimes develop into an antenna, the antenna into a leg, a head into a thoracic segment, a prothoracic segment into a mesothoracic segment and so on. These morphological alterations, by which one structure is changed into another, are called *homoeotic transformations*. They are reasonably common in the insects. However, they do not occur haphazardly but conform to certain specifiable rules: (*a*) dorsal structures will rarely transform to ventral and *vice versa*; (*b*) transformations are predictable; i.e. the organ transformed (autotype) is usually only changed into one or a few other organs (allotypes = organ mimicked by the transformations); (*c*) in *Drosophila* most transformations are into mesothoracic structures. These transformations may be effected by mutation or by teratogenic (development-modifying) agents. The results of the latter mimic the genetic effects and are described as pheno-

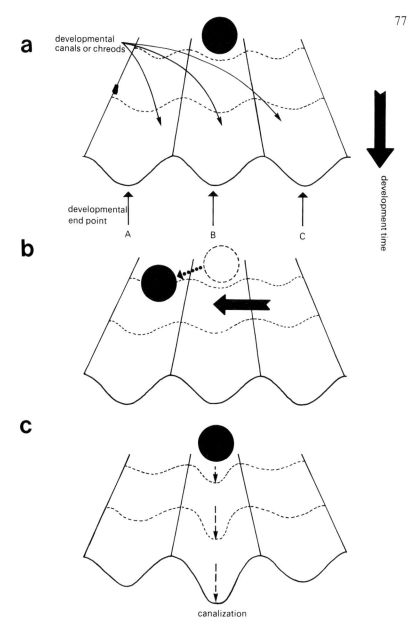

Figure 4.8 Waddington's model of development—conceptualized as a ball moving down a canal or chreod to an end point (*a*). Environmental disturbances (horizontal black arrow) can deflect development to an alternative end point (*b*). Selection can deepen canals and ensure that one end point is favoured (*c*).

copies. These homoeotic transformations illustrate the non-randomness and directionality of morphological transformations. They suggest that there are a restricted number of developmental trajectories available to the organism after genetic or environmental perturbation.

We know very little about the mechanisms behind these various facets of developmental regulation. Waddington (1975) referred to it as *developmental canalization* (the canals = trajectories; Waddington called these chreods) and conceptualized it as a ball moving down a contoured landscape (Fig. 4.8). The ball represents the progress of development. Environmental and genetic disturbances tend to deflect the ball and the extent to which this is allowed depends upon the depth of the canals—

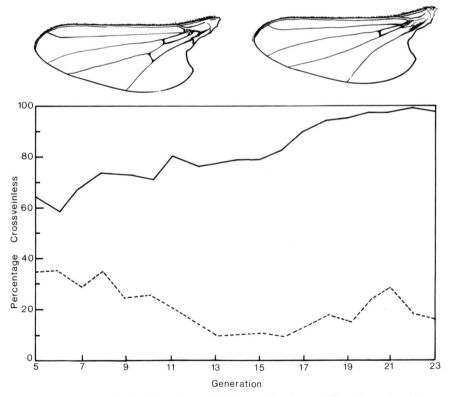

Figure 4.9   Crossveinless wings of *Drosophila* (upper right *cf.* upper left) can be produced by heat shocks. This condition can be made normal by selecting from crossveinless flies after heat shock (as in bottom figure, solid line—broken line is for selection from flies which failed to respond to heat shock).

i.e. degree of canalization. One genotype may allow one of several ontogenetic pathways and the extent to which one is manifest will depend on their relative canalization. An environmental perturbation might push the ball out of one canal into a neighbouring one. At the same time the environmental perturbation can be selecting for the 'deepening' of the alternative canal. This is a potential explanation of the observation that homoeotic transformations can be effected by genetic and/or environmental modifications.

Waddington carried out another interesting investigation on canalization. Under normal conditions, *Drosophila melanogaster* develops a wing with a normal complement of cross veins (Fig. 4.9). Given a 40°C shock during development, the wings lack some of these (i.e. are crossveinless, Fig. 4.9). Moreover, by selecting these aberrant forms and repeating the

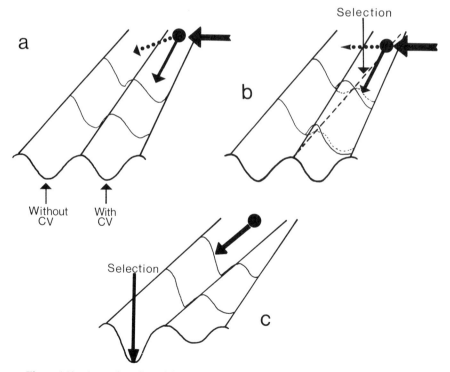

Figure 4.10   An explanation of the results in Figure 4.9. (*a*) Heat shock (black horizontal arrow) deflects the ball to the crossveinless (CV-less) trajectory. (*b*) Selection lowers the contour between trajectories and makes deflection easier. (*c*) The CV-less condition becomes normal and selection deepens (canalizes) this trajectory.

procedure over several generations Waddington discovered that cross-veinless wings would develop automatically without heat shock. Now at first sight this looks like an acquired character being inherited, but there is a more plausible neo-Darwinian explanation and this is illustrated in Figure 4.10. The genetic potential for 'crossveinless' is always there as a 'less preferred trajectory'. The 40°C shock 'flicks' the ball over the hill into this trajectory and selection (in this case artificial) lowers the hill and deepens the alternative canal. The immediate effect is to produce phenocopies and the long-term effect is to produce re-canalization. The same process may have been instrumental in the evolution of hard skin on the human foot. Skin has the genetic potential for 'hardening' given the appropriate stimulus; i.e. abrasion. Skin on the feet is subject to continual abrasion in walking and would be hardened in this way. However, those individuals who were able to respond best to this (i.e. with 'lower hills' and deeper 'hard skin canals') might have been at a selective advantage. Hence the process would be canalized and now the hard skin develops without stimulus even before a baby walks.

4.4 Conclusions

The fact that small genetic adjustments can be amplified by development into larger phenotypic effects, and that developmental processes often operate to resist genetic and environmental disturbances suggests that the relationship between genotype and phenotype is of a non-linear kind. Moreover, the variation in gene-determined traits is restricted and certain changes seem to be more likely than others. Hence, the origin of gene-determined, phenotypic novelty might be said to be directed by the phenotype, and this loosens the concept of randomness as understood in neo-Darwinism. The implications of amplification and canalization for evolutionary theory are as yet only dimly perceived. Clearly, though, we have to take developmental constraints into account within both the adaptationist and genetical programmes of evolutionary biology.

4.5 Further reading

General introductions to the field of 'evolutionary developmental biology' will be found in Waddington (1975), Gould (1977), Alberch et al. (1979) and Alberch (1980). The area is a relatively new and complex one so the reader should not expect to find fully comprehensive treatments in the literature.

CHAPTER FIVE

MACROEVOLUTION

5.1 Introduction

How organisms come to terms with and keep in tune with their environment (sometimes called anagenesis) has been the concern of most of the
preceding chapters. The other important evolutionary process, the genesis
of variety (sometimes called *cladogenesis*), is the issue that is considered in
this final chapter. Macroevolution is a term which refers to the evolutionary
development of taxonomic units at the species level and above. The
evolution of these units is the basis of the generation of taxonomic variety
and so, for the purposes of this book, macroevolution and cladogenesis
will be treated as synonymous. We begin by considering what types of taxa
can be recognized and how they can be classified.

5.2 Taxonomic categories and classification

People have always been interested in naming and classifying the animals
and plants which surround them, and have usually perceived a hierarchy
of taxa in this living world. That is to say, groups of organisms sharing a
few major features in common can be subdivided into groups sharing these
plus other features and these groups can be divided into more subgroups,
and so on. Linnaeus thought this represented a scale of order imprinted on
nature by God and specified a series of categories arranged hierarchically
to define it. His system was:

KINGDOM

CLASS

ORDER

GENUS

SPECIES

Moving down the series, the categories contain units with more and more characters in common. Through the years many additional categories have come into use but the system is still based firmly on the Linnaean hierarchy; most modern biologists recognize the following categories:

KINGDOM

PHYLUM*

CLASS

ORDER

FAMILY

GENUS

SPECIES

* Botanists and microbiologists use the word DIVISION for this category.

All of these may be subdivided further; e.g. at the level of order we might recognize superorders, orders, suborders and infraorders, and these subdivisions can be repeated at each level. Hierarchical classifications of the honey bee and Man are given in Table 5.1.

For Linnaeus, the hierarchical nature of living systems reflected the order due to the creative handiwork of God. For modern biologists it represents an order based upon common descent and evolutionary affinities. The fundamental unit in both schemes is the species—for Linnaeus

Table 5.1 Hierarchical classification of the honey bee and Man

	Honey bee	Man
Phylum	Arthropoda	Chordata
Subphylum		Vertebrata
Class	Insecta	Mammalia
Subclass	Pterygota	Eutheria
Order	Hymenoptera	Primate
Suborder	Apocrita	Simiae
Genus	Apis	Homo
Species	mellifera	sapiens

Note: each species is referred to by using both its genus and species name. Thus:

Man = *Homo sapiens*

Honey bee = *Apis mellifera.*

This is known as a binomial system and was also introduced by Linnaeus. Conventionally, the genus name is started with an upper case letter and the species name with a lower case letter. Both are italicized.

because he believed this was the unit actually formed by God, ¦
evolutionists because this is the unit upon which natural selection
In principle the species can now be objectively defined as a gr(
individuals capable of sharing genes, i.e. of interbreeding, and thus with a
common gene pool available to natural selection, and this is known as the
biological species concept. In practice such groupings cannot be identified
unequivocally, particularly in dead and fossilized individuals, and so a
more pragmatic definition is usually followed—categories into which
taxonomists group individuals. In any event, modern biologists agree that
these groups have evolved from ancestral species such that modern forms
are all more or less closely related, and they are therefore interested in
classifying species into groups which reflect these evolutionary affinities.

Evolution provides two different kinds of event that have to be taken
into account in the construction of classifications: (*a*) order of descent
from ancestors, and (*b*) extent of divergence from ancestors. In general, the
former emphasizes similarities and the latter differences. Imagine three
species arising at the same time from a common ancestor. Two develop
only minor differences, whereas the third diverges rather markedly from
the other two (e.g. evolves a key innovation). If we classified on the basis
of (*a*) we would have to view the three species as equally related, whereas
using (*b*) leads to a separation of the species with the innovation from the
other two. There are three views on how much weight should be placed on
these criteria.

(1) *Cladistics* (Hennig, 1979; Eldredge and Cracraft, 1980)

This derives from principles formulated by a German entomologist,
W. Hennig, in 1950 (but see Hennig, 1979), and lays emphasis on
genealogy, i.e. the branching sequences, as being the only objective way of
taxonomic ordering. ('Cladistics' is derived from the Greek *klados* = branch
or young shoot.) All members of a taxon (clade) should stem from the
same ancestor, and organisms should be grouped *only* according to the
sequence of separation from a common ancestor. In this analysis it is
usually assumed that all the species present at any one time are related by
descent from ancestral forms (Fig. 5.1*a*, i not ii), and are not themselves
ancestors of any other extant species. Hence all contemporary species can
be objectively ordered in a sequence based strictly on genealogical affinity,
and such an ordering can be represented diagrammatically as a *cladogram*
(Fig. 5.1*a*, iii). These cladograms are constructed on the basis of the
observed, or presumed, number of characters shared in common by

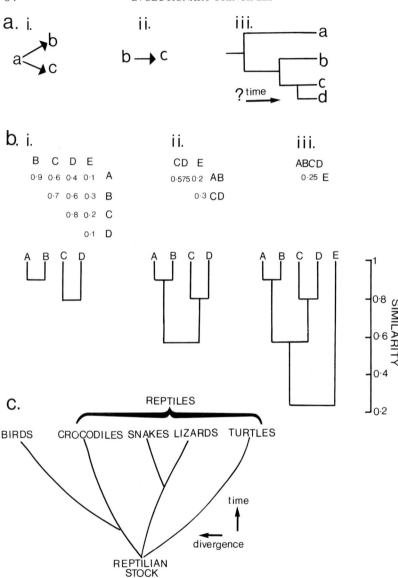

Figure 5.1   (*a*) Cladists usually assume that species under consideration are related as in (i) not as in (ii); (iii) is a cladogram. The latter is based on branching sequences. The question mark associated with the time axis is because there are at least two interpretations of cladograms. One assumes that they are evolutionary trees in which ancestors are not specified, so time is included (classical cladistics), and the other assumes that they are

groups. However, similarity *per se* has to be treated with caution, since it might only reflect convergence (different structures evolving a similar appearance because of a common objective—e.g. the wings of bats, birds and insects). Evolutionary relationships can only be based on characters that are similar because they are inherited from a common ancestor, and these are called *homologies*. The more homologies that two species share, the more closely they are related. Two species that are closest relatives share homologies that are not shared by any other species. However, we would not get very far by classifying on the basis of fundamental characters basic to most organisms (e.g. multicellularity, or even the pentadactyl organization of the vertebrate limb) because, by definition, these *are* possessed by most organisms. These are called *shared, primitive characters* (sometimes called symplesiomorphies). Cladists argue that branching sequences can only be recognized by identifying *shared, derived characters* (sometimes called synapomorphies)—homologous similarities present at the branching point itself. Take, for example, a cat, a lion, a seal and human. These all have many shared, primitive characters—e.g. multi-cellularity, blood, pentadactyl limbs, etc. However, a large brain and binocular vision are shared, derived characters unique to primates and mark off the human as a first branching point. The cats and lions have more shared, derived characters (e.g. loss of upper molars, conversion of upper premolar to a cutting blade or carnassial tooth) than either does with the seal, and so this defines our second branching point. The kind of cladogram which emerges from the analysis is illustrated in Figure 5.1*a*, iii, where a = human; b = seal; c = lion; d = cat.

---

patterns of relationship with branching points representing synapomorphies so time is excluded (transformed cladistics). Classical cladistics is interested in *evolutionary processes* and transformed cladistics in *taxonomic patterns*. Clearly, the latter should relate to the former, but some cladists would prefer to remain neutral on the exact nature of the relationship. (*b*) Stepwise construction of a phenogram. Letters represent species. Numbers in matrices are coefficients of overall similarity (ranging from 0 = no similarity to 1 = complete similarity) and are based on all characters measured. In step (i) the matrix contains similarity coefficients among the five species. We begin to construct the phenogram by searching for basic pairs—i.e. similarity coefficients are maximal for each member of the pair. The basic pairs are joined at appropriate levels on an axis representing the similarity coefficients. Step (ii) considers each basic pair as a single unit, and similarly coefficients here are averages of all individual species, i.e. $0.575 = (0.6+0.4+0.7+0.6)/4$. Again we search for basic pairs and make appropriate connections on the graph. Step (iii) continues and completes this procedure. After Luria, S. E., Gould, S. J. and Singer, S. (1981) *A View of Life*. Benjamin/Cummings, Menlo Park, California. (*c*) Phylogenetic tree which tries to take both branching sequences (and hence timing) and the extent of divergence into account. The former is usually represented as precisely as possible on the vertical axis; the latter is usually based on subjective assessments and is represented imprecisely on the horizontal axis.

One final question: How does the cladist distinguish between primitive and derived characters? An important technique is the so-called *out-group comparison*. If the similarities between species within the group being considered are also found in species outside the group, then they are more likely to be primitive than derived. Potentially, a vast array of species can be compared, and this is essentially what we do unconsciously when labelling traits like multicellularity as being primitive. However, it usually makes practical sense to place some limits on this kind of study, and intuitively the most sensible approach is to compare species within the group to other *similar ones* outside the group.

(2) *Phenetics* (Sokal and Sneath, 1963; Sneath and Sokal, 1973)

Biologists of this school suggest that it is impossible to classify organisms objectively according to genealogy, principally because it is impossible to definitely distinguish between homologies and convergent characters. Instead they focus attention on all similarities, and attempt to classify organisms objectively on the basis of these (i.e. on the extent of overall similarity). They claim that if sufficient numbers of characters are taken into account the homologies should overwhelm the convergent characters. However, there is no *a-priori* guarantee of this. Other pheneticists claim that all they want to produce are classifications of convenience, such as the ones used to classify books in libraries, not necessarily ones which reflect evolutionary relationships. Sophisticated mathematical tools are often used to effect phenetic analysis (i.e. assess the extent of overall similarity) and the results are usually summarized as phenograms (Fig. 5.1*b*).

(3) *Evolutionary systematics* (Mayr, 1970)

The classifications here are based upon a *combination* of genealogy and the extent of overall similarities and differences reflecting the extent of divergence. There are no objective rules for doing this and much rests on the experience of taxonomists in assessing the relative importance of the various criteria. They summarize their results as phylogenetic trees (Fig. 5.1*c*).

The following example, involving birds and reptiles, illustrates some of the differences between these major schools. Both birds and crocodiles are derived from a common archosaurian ancestral group that included the dinosaurs. The common ancestor of turtles, lizards and snakes, crocodiles and birds was a more distant reptile. Hence birds and crocodiles share more characters in common than the crocodile does with other extant reptiles.

Yet, after separation, the birds have evolved more rapidly than the crocodiles and now have very distinctive features—ability to fly, feathers, homeothermy, etc. Since the birds have diverged further from the reptilian stock than the crocodiles, the evolutionary systematist separates them off from all the other reptiles, grouping them in a separate class—Aves—and groups all the extant reptiles, including crocodiles, in a single class— the Reptilia (Fig. 5.1c). Phenetics would generate a similar grouping on the basis of morphological similarity. Alternatively, the cladistic school classifies strictly on the basis of genealogical affinity and hence generates a cladogram in which birds and crocodiles are more closely related than the latter are to turtles, snakes and lizards.

To summarize: there is probably no best way to construct biological classifications that reflect evolutionary relations. Cladists do it objectively on the basis of genealogy only, but ignore important information on divergence. Pheneticists do it objectively on the basis of similarities but inevitably confuse homologies and convergent characters. Evolutionary systematists try to take into account both genealogy and divergence but cannot do this entirely objectively.

Irrespective of which methodology is followed, however, once the nested groups of taxa have been distinguished, the next step is to fit them into the Linnaean hierarchy—i.e. assigning a definite rank, such as order or family, to them. Cladists often give a new rank after each branching point in the cladogram. The evolutionary systematist, however, ranks taxa by the degree of divergence from the common ancestor, often assigning different ranks to sister groups. This process is extremely subjective and therefore rankings are open to much debate and subject to much alteration and adjustment.

5.3   How taxa are formed—the neo-Darwinian view or Modern Synthesis

The so-called Modern Synthesis tries to explain the genesis of taxonomic variety, i.e. macroevolutionary phenomena, in terms of the principles of neo-Darwinism, i.e. microevolutionary processes. Important landmarks in this approach are the books of Dobzhansky (1937), Huxley (1942), Mayr (1942) and Simpson (1944).

The Modern Synthesis bases its explanation of speciation on the fundamental principles of neo-Darwinism, namely:

1. point mutations, particularly in structural genes, are the source of variation;
2. evolutionary change is a result of adjustments in gene frequencies;
3. the direction of these adjustments is determined by natural selection.

Thus classically, populations of a species are supposed to become physically isolated and there is then no gene flow between them. This isolation produces groups which are more or less random samples from the original gene pool. Non-random sampling is possible, e.g. as occurs in the founder effect, but is rare according to Synthesists. The isolates become adapted to the conditions in which they are isolated, so their gene frequencies move apart and the populations ultimately become so different that even if the isolating barriers are removed there can be no inter-breeding—i.e., they are now true species (see 5.2). Such a process of speciation depends crucially on an initial isolating event and then involves slow and gradual changes. It is illustrated schematically in Figure 5.2.

There are two problems associated with this gradualistic interpretation; one of fact and the other of logic. On the facts, the fossil record often does not display gradual changes. Many species remain apparently unchanged for millions of years and then abruptly disappear, to be replaced with something that is substantially different but clearly related to the original. A good example is given by the work of Williams (1981) on a group of

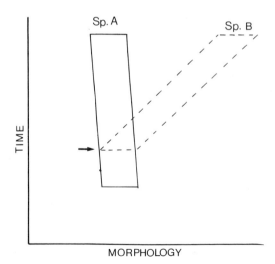

Figure 5.2 Gradualistic interpretation of speciation. After isolation (horizontal arrow), the isolate (broken line) *gradually* diverges from the 'parent' species (A).

freshwater molluscs from late Cenozoic deposits of Lake Turkana in northern Kenya. Over the last 5 million years, population measures show only small changes in 16 of the species considered. On the other hand, five species change little for most of the time but then very rapidly over periods of approx. 50 000 years. On the logic, there is what Mivart referred to as the dilemma of the incipient stages of useful structures. If the evolution of structures occurs gradually, what value are structures with intermediate characters? For example, what use is a partially formed wing to a reptile (see also Section 1.5)?

The Synthesists (or Gradualists) have addressed the factual problem by arguing that it does not occur as frequently as is sometimes suggested and, when it does, that it can be explained in terms of the imperfections of the fossil record. The accumulation of sediments and the entrapment of bones is a capricious process, particularly for groups which exist at low population densities, and intermediate forms *are* likely to be rare. Mivart's dilemma has been explained by invoking *preadaptation*. The intermediate structures act in a particular way for which they are well suited, but by chance they are appropriate for other roles which they are able to play after further elaboration. A flap of skin between the fore-limb and the trunk might have evolved in reptiles as a thermoregulating device and then have been found of use in gliding away from enemies, ultimately giving rise to a wing. Similarly, the air-bladder of fishes was not evolved so that they might one day invade land but as an accessory respiratory organ for use in the aquatic environment. However, once evolved, air-bladders were of use as lungs when new environmental problems, such as the drying of ponds, appeared.

5.3.1 *A note on isolation and speciation*

It was stressed above that isolation was an essential prerequisite of speciation. However, it is now recognized that there are several possible ways that the isolation might occur and hence the speciation might be effected. These are reviewed by Bush (1975).

Gene flow (the sharing of genes) between populations might be suppressed by a physical or geographical barrier. This process is called *allopatric speciation*. Alternatively, in principle, gene flow can be suppressed in populations which exist *in the same area*—if, for example, the bearers of favourable mutations mate preferentially with each other and if the selection of these mutations is sufficiently intense to overcome any gene flow from matings with more common parental forms. This is *sympatric speciation.* Finally, even if there is some gene flow between partially isolated populations in adjacent areas there can, again in principle, be a divergence of characters under the influence of selection if gene flow is relatively low and the differences in selection pressure are relatively high. This is known as *parapatric speciation* and is likely to be common in organisms with low vagility (= dispersal powers) such as plants and land snails.

Whereas few biologists doubt the importance of the allopatric process in evolution, there is much debate on the importance of the sympatric and parapatric processes. The importance of deme structures (e.g. as in Fig. 2.9*E*) in the process of speciation is, however, again underlined.

A distinction should also be noted between geographic isolation (which is what was discussed under the heading of allopatry) and isolating mechanisms. The latter are the character differences between populations which prevent interbreeding and maintain the integrity of species when they occur together. These are many and varied and may involve *premating isolating mechanisms* (e.g. morphological or behavioural characters which prevent males and females copulating) and *postmating isolating mechanisms* (e.g. embryological incompatibilities which lead to infertile eggs, or sterile offspring, for example mules).

5.4   An alternative, punctuational view

There is an alternative to the gradualistic position outlined in the preceding section. The fossil record might not be as imperfect as has been suspected. Evolution might actually occur by short punctuations of rapid change followed by long periods of equilibrium. This punctuated equilibrium theory has the advantage of acknowledging what is apparently observed, and it solves Mivart's dilemma, for the intermediate stages might never have existed. This implies that larger changes in form and function can take place rapidly. Darwin and the neo-Darwinists thought that such a view would undermine the credibility of their theories because the vast majority of large changes should be fatal; i.e. incompatible with the phenotype (see 1.5). Actually, this is also in line with the Punctuationalists' views, for the long periods of stasis are thought by this school to reflect the inability of the majority of mutational changes to integrate with the phenotype; i.e. constraints associated with coadaptation and integration resist change. Occasionally, however, one of these changes, what Goldschmidt referred to in his book *The Material Basis of Evolution* (1940, Yale University Press) as a hopeful monster, might turn out to be

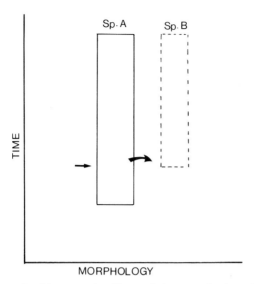

Figure 5.3 Punctuational interpretation. The speciation event (horizontal arrow) is rapid, but thereafter there is little change.

compatible with the existing adaptive complex. The origin and integration of such large, mutational changes is now known to be consistent with developmental processes (Chapter 4). Alternatively, a dramatic shift might be achieved away from the gene pool of the parent population by a non-random sampling in the initial stages of isolation. This is thought to be particularly likely in small, *peripheral isolates* of large populations. The resulting punctuational interpretation is depicted in Figure 5.3. (Note the similarity between this figure and Figure 5.1a, iii; i.e. the cladogram. Cladists are often very sympathetic to punctuated equilibrium theory, and punctuationalists often espouse cladism. This is because cladistic analysis assumes rapid speciation events with little divergence between the branching points.)

As with traditional neo-Darwinism, therefore, punctuated equilibrium theory perceives the initial source of variation as being random relative to the course of evolution, but sees this variation as involving changes of much greater magnitude. Moreover, because of this and the subsequent stasis, it envisages the source of variation, i.e. the isolations or speciation events, as being the most important element in determining the course and rate of evolution (a return to the view of the early Mendelists?).

But if this is so, from what processes do adaptations arise and how are

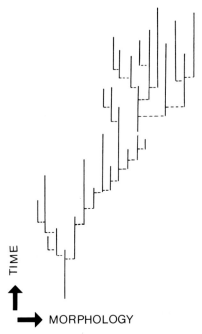

Figure 5.4   A trend produced by differential origin and survival of species, not by gradual changes. Species arise suddenly and then show little change. The shift to the right occurs because more speciation events occur in that direction, and because species with these morphologies survive longer. It is conceivable that either of the processes, differential speciation and extinction, might predominate.

evolutionary trends generated—given that both of these are impressive features of the fossil record and of extant species? The answer, for the Punctuationalists, is *species selection* (a term coined by Stanley, 1975). According to this thesis, trends may occur (1) because some randomly generated isolates are better at persisting than others, either (*a*) because they are better adapted in the conventional sense or (*b*) because of the ecological conditions in which they are isolated (e.g. chance absence of a voracious predator); and/or (2) because some species, as a result of their organization or ecology, speciate more frequently than others. These possibilities are illustrated further in Figure 5.4. Both (1) and (2) generate trends, but the adaptations that arise, particularly as a result of (1*b*) and (2), are not likely to be as precise as those envisaged as a result of the more conventional gradual speciation.

In its modern form this theory of punctuated equilibrium has been advocated particularly by Gould and Eldredge (1977) and by Stanley (1975, 1979). However, as already noted, it has its roots in the writings of Goldschmidt (see above) and also De Vries (1905). T. H. Huxley, one of Darwin's staunchest allies, was also inclined to advocate considerable *saltus* in evolution—much to Darwin's own disgust!

5.5 Comparison of gradualism and punctuationalism

To make more concrete the difference between punctuationalism and gradualism in terms of evolutionary trends and adaptations, consider the classical example of the evolution of the modern horse, whose distant ancestor *Hyracotherium* was a three-toed creature, no bigger than a dog. The fossil record shows an apparently gradual change in body size and form leading eventually to the familiar *Equus*. The classical gradualistic account is in terms of a progressive change under the influence of natural selection (see 3.2). By contrast, the punctuated equilibrium theory explains trends in horse evolution as the result of a differentially pruned bush. The evolutionary history of the horse is considered as a speciating lineage with some new species projecting in the direction of bigger bodies and fewer toes but others involving completely different changes in form. The ones with the bigger bodies and reduced number of digits thrive more success-fully than those with the more primitive features of more toes and smaller bodies, and this produces an asymmetry in the evolutionary bush. The 'centre of gravity' through time leans steadily towards the bigger species with one toe (e.g. as in Fig. 5.4).

Table 5.2 lists the major features of difference between the gradualist and punctuationalist theories.

Table 5.2 A comparison of the gradualist and punctuational hypotheses

	Gradualist	*Punctuational*
Isolation	random process, occasionally biased, usually allopatric	subgroup isolated often with biased genetic composition, can be sympatric or parapatric
Source of variation	mainly point mutations	mainly macromutations (i.e. mutations in original sense)
Selection	within populations	between populations (or species)
Divergence	slow and continuous (and proportional to number of generations)	rapid (not proportional to number of generations) but with long periods with no change
Constraints	not very important	very important
Adaptations	obvious and tight	not so obvious and loose

5.6   Tests of the two theories

Stanley (1975, 1979) reviews various kinds of evidence which can be used to test between the gradualist and punctuationalist theories. Amongst the most general, however, is the expectation that the rate of macroevolution (i.e. the rate of formation of families and orders which reflect major morphological changes) should (1) on the gradualistic model be proportional to the *time* or, more properly, *number of generations* available for diversification and (2) on the punctuational model be proportional to the amount of splitting, i.e. speciation, that has taken place. If these two parameters, time and rate of splitting, can be separated, then in principle

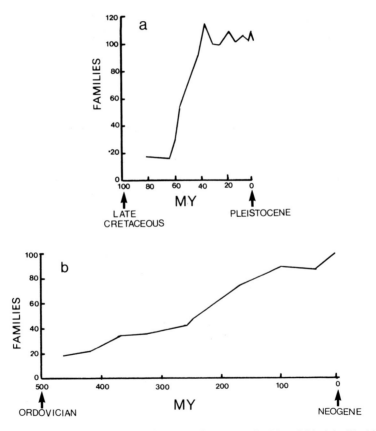

Figure 5.5 Diversity of families through time for Mammalia (*a*) and Bivalvia (*b*). After Stanley, S. M. (1979) *Macroevolution*. W. H. Freeman & Co., San Francisco.

we can distinguish between the two hypotheses. On this basis, Stanley points to the comparison between mammals and bivalve molluscs. In the former, the rate of speciation has been rapid and the major radiation within the group, to approximately 100 families, has occurred over less than 30 My (million years). Alternatively, the speciation of the bivalves has been slow and there has been a correspondingly slow radiation over more than 300 My (Fig. 5.5a, b). Moreover, within the bivalves one group of cockles, the so-called Pontian series, which became isolated from the Mediterranean in the Caspian Sea, has speciated more rapidly than other close relatives and within less than 3 My formed such a diverse set of morphologies that the group is now divided into four sub-families.

The corollary of these arguments is that in groups where there has been little evolutionary change over long periods of geological time there should have been little speciation. And, indeed, this seems to have been the case for many of the so-called living fossils. The lungfishes fit this prediction very well. They underwent rapid changes in morphology only when speciating rapidly in the mid-Paleozoic, soon after appearing. Thereafter the rate of speciation decreased abruptly and so did the rate of morphological change. Similarly, alligators, which have shown little structural evolution, include only two living species and have at no time harboured a large number of species. Finally, there is but a single living species of aardvark, and the fossil record reveals only a small number of lineages as far back as the base of the Miocene where the earliest-known representatives are found.

All this argues for the punctuationalist position. However, the evidence is not decisive, for this analysis depends crucially on the separation of fossil taxa which might reflect the predilections of taxonomists more than the effects of evolution; e.g. the question of which groups of species constitute families is rather subjective. Even if the data are accepted at face value, the traditional neo-Darwinist might still claim that a number of factors, and particularly variation in the intensity of natural selection, overwhelm the expected correlation between macroevolutionary rate and time predicted in (1) above. In other words, neo-Darwinism does not necessarily predict gradualism. There might be rapid periods of change under the influence of intense selection and slow periods under the influence of weak selection. Moreover, stasis in hard parts, which are the ones preserved in the fossil record, need not mean that there is stasis in soft structures, and metabolic and behavioural processes. The evidence, then, though apparently supporting the punctuationalist theory, is still far from conclusive and so the issue is still a live one, hotly debated by both sides.

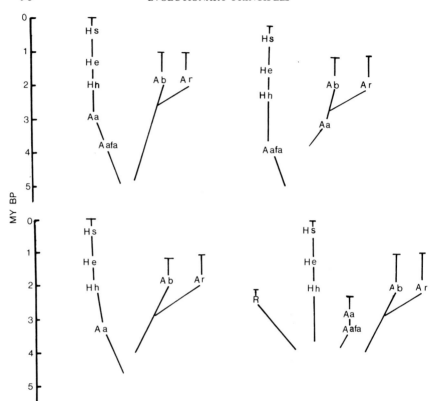

Figure 5.6   Alternative phylogenies proposed for the major hominid taxa. *R = Ramapithecus;*
Aafa = Australopithecus afarensis; Aa = *A. africanus;* Ab = *A. bosei;* Ar = *A. robustus;*
Hh = *Homo habilis;* He = *H. erectus;* Hs = *H. sapiens.* After Cronin, J. E. *et al.* (1981)
Nature, **292,** 113–122. BP = before present.

5.7   Human evolution—gradual or punctuated?

The creationists (Chapter 1) thought that Man was sharply distinct from
the apes—a divine punctuation had interposed the soul! As a reaction to
this, the Darwinists and neo-Darwinists have tenaciously held to a
gradualistic interpretation of the progressive evolution of Man from the
apes. On this basis it has been standard practice to put *Homo sapiens* at
the end of a continuous series tracing back through at least two other
species; e.g. Figure 5.6. The various taxa have been characterized as
follows.

(1) *Ramapithecus* (from Rama = Indian god, pithekos = Greek for ape). This group lived approximately 8–14 My ago. They are known only from jaw fragments, but these suggest a flattened face with a short muzzle.

(2) *Australopithecus* (austral = Latin for southern). This group lived 1.5–5 My ago. They had a flattened face. The orientation of the skull relative to the spinal cord, and the bones of the hind-limbs relative to the pelvis, suggest an upright gait. There are two main groups: *Robust* australopithecines (*A. robustus*), which had a heavy skull and dentition suggesting a vegetarian diet, and *Gracile* australopithecines (*A. afarensis* and *A. africanus*) with more delicate skeletons.

(3) *Homo habilis* ('handy man'). This group lived 1.5–2 My ago. They had an erect gait, relatively large brains and were tool-users.

(4) *Homo erectus* ('upright man'). This group lived 1 My ago. They had a gait even more like ours and larger brains than the habilines, and they used fire.

(5) *Neandertals* (sometimes Neanderthals, from the German for Neander Valley, where they were first discovered). This group lived around 0.5 My

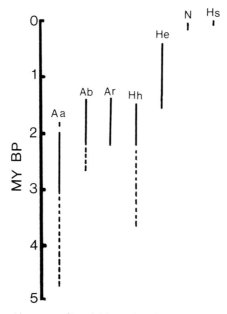

Figure 5.7 Stratigraphic ranges of hominids. Broken lines represent fossil fragments tentatively assigned to species. Abbreviation as in Figure 5.6, and N = Neanderthal. After Stanley, S. M. (1979) *Macroevolution.* W. H. Freeman & Co., San Francisco.

ago, and are so like modern man, in morphology and in having a definite culture with ceremonies, that they are classified as a subspecies of *Homo sapiens*.

Recently two important pieces of evidence have been mustered that resurrect the view that human evolution is punctuated (Stanley, 1979). First, it has been suggested that more than one species of hominid existed at one time, indicating a bush-like rather than linear phylogeny within the genus *Homo*. Second, it is suggested that there is evidence for relatively long periods of stasis within both the australopithecine and hominid groups (e.g. see *Australopithecus africanus* and *Homo habilis* and *Homo erectus* in Figure 5.7) and this is rather startling considering the high net rate of evolution which is characteristic of human evolution.

On the other hand, the evidence for a temporal overlap in the species of genus *Homo* is not undisputed and, even if it were correct, does not automatically negate the gradualistic interpretation. The branching pattern might simply have increased the number of directions along which human evolution gradually proceeded. Moreover, and more crucially, though

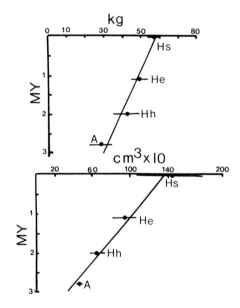

Figure 5.8   Changes in mean body weight (upper) and cranial capacity (lower) with time in hominids. Abbreviations as in Figure 5.6. After Cronin, J. E. *et al.* (1981) *Nature*, **292**, 113–122.

certain morphological traits, like those of the skulls, might show sharp distinctions between the various hominid taxa, the changes in quantitative characters, i.e. body size and cranial capacities, show a nearly continuous progression (Fig. 5.8). It is also claimed by some anthropologists that the stasis within taxa is more apparent than real (Cronin *et al.*, 1981). For example, the earliest examples of *H. erectus* had a longer and flatter face, shorter, thinner and less buttressed cranial vault, and smaller cranial capacity than later *H. erectus*.

5.8 Conclusions

There is still much room for debate between punctuationalists and traditional neo-Darwinists. The main problem facing the latter is probably not the rapidity with which evolutionary changes can occur during the so-called punctuations, for given appropriate intensities of selection rapid changes can be effected, but the fact that traits can remain static over long periods of geological time. For example, Lande (1976) calculated that the number of selective deaths necessary to account for the change in certain skeletal structures in Cenozoic mammals was approximately 1 per million individuals. Such selection pressures are so weak that there is no reason to exclude the possibility that they were generated by completely random processes. And yet wholesale genetic drift, occurring over millions of years, is quite outside the compass of neo-Darwinism. Alternatively, the neo-Darwinists might argue that behind the apparent stasis there is intense stabilizing selection (see Section 2.3.3), but then we have to posit long-term environmental stasis which seems equally unlikely.

The main problem facing punctuationalism, on the other hand, is the question of randomness in the saltational changes. Punctuational changes have to be equally likely in all directions, including that of the evolutionary trend. This could occur through drift and founder effects, but the widespread occurrence of these processes has yet to be demonstrated. Alternatively, if large mutations are the basis of the punctuation they must confer an advantage on their bearers to spread rapidly, and this is only distinguishable from classical neo-Darwinism in terms of the magnitude of the change.

Whether these issues can be settled clearly by reference to data remains doubtful. Moreover, it is conceivable that the evolution of some groups might fit the punctuational model and others the neo-Darwinian one. In other words it is unlikely that one evolutionary model will be universally applicable.

5.9   Further reading

A good history of the Modern Synthesis is to be found in Mayr and
Provine (1980). To appreciate how different evolutionary biologists have
interpreted the punctuated equilibrium theory, the reader should compare
and contrast accounts of the same conference on macroevolution (in
Chicago, October 1980) by Lewin (1980) and Maynard Smith (1981). A
New Scientist issue, marking the centenary anniversary of the death of
Charles Darwin (15 Apr. 1982, **194**, no. 1301), contains arguments for the
punctuational position by Gould, and against by Charlesworth, and also
a good article on human evolution by Stringer. A popular account of the
punctuated equilibrium model is given in Stanley (1982).

REFERENCES

Chapter 1

George, W. (1982). *Darwin.* Fontana, London.

Gillespie, N. C. (1979). *Charles Darwin and the Problem of Creation.* University of Chicago Press, Chicago and London.

Howard, J. (1982). *Darwin.* Oxford University Press, Oxford.

Moore, J. R. (1979). *The Post-Darwinian Controversies.* Cambridge University Press, Cambridge.

Moorehead, A. (1969). *Darwin and the 'Beagle'.* Hamilton, London.

Ridley, M. (1982). Coadaptation and the inadequacy of natural selection. *British Journal of History of Science,* **15**, 45–68.

Ruse, M. (1979). *The Darwinian Revolution.* University of Chicago Press, Chicago and London.

Ruse, M. (1982). *Darwinism Defended.* Addison-Wesley Publ. Co., Massachusetts.

Smith, C. U. M. (1976). *The Problem of Life.* Macmillan Press Ltd., London.

Chapter 2

Ayala, F. J. and Kiger, J. A. (1980). *Modern Genetics.* Benjamin/Cummings Publ. Co., Menlo Park, California.

Berry, R. J. (1982). *Neo-Darwinism.* Edward Arnold, London.

Charlesworth, B. (1980). *Evolution in Age-structured Populations.* Cambridge University Press, Cambridge.

Cook, L. M. (1971). *Coefficients of Natural Selection.* Hutchinson University Library, London.

Crow, J. F. (1979). Genes that violate Mendel's rules. *Scientific American,* **240**, 104–113.

Falconer, D. S. (1981). *Introduction to Quantitative Genetics,* 2nd edn. Longman, London and New York (1st edn. Oliver & Boyd, 1960).

Gale, J. S. (1980). *Population Genetics.* Blackie, Glasgow and London.

Ludovici, L. J. (1963). *The Chain of Life.* Phoenix House Ltd., London.

Ricklefs, R. E. (1980). *Ecology,* 2nd edn. Nelson, Middlesex.

Roughgarden, J. (1979). *Theory of Population Genetics and Evolutionary Ecology: An Introduction.* Macmillan Publ. Co., N.Y.

Turner, J. R. G. (1970). Changes in mean fitness under natural selection. In *Mathematical Topics in Population Genetics* (ed. K. Kojima) Springer, Berlin, pp. 33–78.

Watson, J. D. (1976). *Molecular Biology of the Gene,* 3rd edn. W. A. Benjamin Inc., California.

Watson, J. D. (1980). *The Double Helix: A Norton Critical Edition* (ed. G. S. Stent) W. W. Norton, New York.

Watson, J. D. and Tooze, J. (1981). *The DNA Story*. W. H. Freeman & Co., San Francisco.

Woods, R. A. (1980). *Biochemical Genetics*. Chapman & Hall, London and New York.

Zimmering, S., Sandler, L. and Nicoletti, B. (1970). Mechanisms of meiotic drive. *Ann. Rev. Genetics*, **4**, 409–436.

Chapter 3

Alexander, R. McN. (1982). *Optima for Animals*. Edward Arnold, London.

Bell, G. (1982). *The Masterpiece of Nature*. Croom Helm Publishers, London.

Blower, G. (1969). Age-structure of millipede populations in relation to activity and dispersion. In *The Soil Ecosystem* (ed. J. G. Sheal) Systematics Association Publ. No. 8, pp. 209–216.

Cain, A. J. and Sheppard, P. M. (1954). Natural selection in *Cepaea*. *Genetics*, **39**, 89–116.

Calow, P. and Townsend, C. R. (1981). Energetics, ecology and evolution. In *Physiological Ecology. An Evolutionary Approach to Resource Use* (eds. C. R. Townsend and P. Calow) Blackwell Scientific Publications, Oxford, pp. 3–19.

Calow, P. and Woollhead, A. S. (1977). The relation between ration, reproductive effort and age-specific mortality in the evolution of life-history strategies—some observations on freshwater triclads. *J. Anim. Ecol.*, **46**, 765–781.

Cohn, D. L. (1954). Optimal systems. I The vascular system. *Bull. Math. Biophys.*, **16**, 59–74.

Cohn, D. L. (1955). Optimal systems. II The vascular system. *Bull. Math. Biophys.*, **17**, 219–223.

Cole, L. C. (1954). The population consequences of life history phenomena. *Quart. Rev. Biol.*, **29**, 103–137.

Davies, P. S. (1966). Physiological ecology of *Patella* I. The effect of body size and temperature on metabolic rate. *J. Marine Biol. Ass., U.K.*, **46**, 647–658.

Davies, P. S. (1969). Physiological ecology of *Patella* III. Desiccation effects. *J. Marine Biol. Ass., U.K.*, **49**, 291–304.

Dawkins, R. (1976). *The Selfish Gene*. Oxford University Press, Oxford.

Dawkins, R. (1982). *The Extended Phenotype*. W. H. Freeman and Co., Oxford and San Francisco.

Dobzhansky, Th. (1937). *Genetics and the Origin of Species*. Columbia University Press, N.Y.

Fisher, R. A. (1930). *The Genetical Theory of Natural Selection*. Clarendon Press, Oxford.

Ford, E. B. (1975). *Ecological Genetics*. Chapman & Hall, London.

Gilpin, M. E. (1975). *Group Selection in Predator-Prey Communities*. Princeton University Press, Princeton, N.J.

Gould, S. J. and Lewontin, R. C. (1979). The spandrels of San Marco and the Panglossian paradigm: a critique of the adaptationist programme. *Proc. Roy. Soc. Lond.*, **205B**, 581–598.

Hamilton, W. D. (1964). The genetical evolution of social behaviour, I and II. *J. Theoret. Biol.*, **7**, 1–32.

Hamilton, W. D. (1972). Altruism and related phenomena, mainly in social insects. *Ann. Rev. Ecol. Syst.*, **3**, 193–232.

Kettlewell, H. B. D. (1956). Selection experiments on industrial melanism in the Lepidoptera. *Heredity*, **10**, 14–22.

Kettlewell, H. B. D. (1973). *The Evolution of Melanism*. Clarendon Press, Oxford.

Krebs, J. R. and Davies, N. B. (1981). *An Introduction to Behavioural Ecology*. Blackwell Scientific Publications, Oxford.

Lewontin, R. C. (1970). The units of selection. *Ann. Rev. Ecol. Syst.*, **1**, 1–18.

Lewontin, R. C. (1978). Adaptation. *Scientific American*, **239**, 156–165.

Maynard Smith, J. (1972). *On Evolution*. Edinburgh University Press, Edinburgh.

Maynard Smith, J. (1976). Evolution and the theory of games. *American Scientist*, **64**, 41–45.

Maynard Smith, J. (1978). *The Evolution of Sex*. Cambridge University Press, Cambridge.

Maynard Smith, J. (1982). *Evolution and the Theory of Games*. Cambridge University Press, Cambridge.

Metcalf, R. A. (1980). Sex ratios, parent-offspring conflict, and local competition for mates in the social wasp *Polistes metricus* and *Polistes variatus*. *American Naturalist*, **116**, 642–654.

Milsum, J. H. and Roberge, F. A. (1973). Physiological regulation and control. In *Foundations of Mathematical Biology*, **3** (ed. R. Rosen) Academic Press, London and New York, pp. 1–95.

Muller, H. J. (1954). The relation of recombination to mutational advance. *Mut. Res.*, **1**, 2–9.

Parker, G. A., Baker, R. R. and Smith, V. G. F. (1972). The origin and evolution of gamete dimorphism and the male–female dichotomy. *J. Theoret. Biol.*, **36**, 529–553.

Rosen, R. (1967). *Optimality Principles in Biology*. Butterworths, London.

Sahlins, M. (1976). *The Use and Abuse of Biology: An Anthropological Critique of Sociobiology*. University of Michigan Press, Ann Arbor, Michigan.

Shorrocks, B. (1978). *The Genesis of Diversity*. Hodder & Stoughton, London.

Stearns, S. C. (1976). Life-history tactics: a review of ideas. *Quart. Rev. Biol.*, **51**, 3–47.

Stearns, S. C. (1977). The evolution of life-history traits: a critique of the theory and a review of the data. *Ann. Rev. Ecol. Syst.*, **8**, 145–171.

Sved, J. A. and Mayo, O. (1970). The evolution of dominance. In *Mathematical Topics in Population Genetics* (ed. K. Kojima) Springer Verlag, New York, pp. 289–316.

Townsend, C. R. and Calow, P. (eds.) (1981). *Physiological Ecology: An evolutionary approach to resource use*. Blackwell Scientific Publications, Oxford.

Turner, J. R. G. (1967). Why does the genome not congeal? *Evolution*, **21**, 645–656.

Williams, G. C. (1975). *Sex and Evolution*. Princeton University Press, Princeton.

Wilson, D. S. (1980). *The Natural Selection of Populations and Communities*. Benjamin/Cummings, Menlo Park, California.

Wilson, E. O. (1975). *Sociobiology: the New Synthesis*. Harvard University Press, Harvard.

Wilson, E. O. and Bossert, W. H. (1971). *A Primer of Population Biology*. Sinauer Associates, Stanford, Connecticut.

Chapter 4

Alberch, P. (1980). Ontogenesis and morphological diversification. *American Zoologist*, **20**, 653–667.

Alberch, P., Gould, S. J., Oster, G. F. and Wake, D. B. (1979). Size and shape in ontogeny and phylogeny. *Paleobiology*, **5**, 296–317.

Bard, J. B. (1977). A unity underlying the different zebra striping patterns. *J. Zool.*, London, **183**, 527–539.

Dobzhansky, T. (1937). *Genetics and the Origin of Species*. Columbia University Press, New York.

Edmonds, H. W. and Sawin, P. B. (1936). Variations of the branches of the aortic arch in rabbits. *American Naturalist*, **70**, 65–66.

Gould, S. J. (1977). *Ontogeny and Phylogeny*. Harvard University Press, Cambridge, Mass.

Huxley, J. S. (1932). *Problems of Relative Growth*. Methuen, London.

Newell, N. D. (1949). Phyletic size increase, an important trend illustrated by fossil invertebrates. *Evolution*, **3**, 103–124.

Needham, J. (1933). On the dissociability of the fundamental processes in ontogenesis. *Biol. Revs.*, **8**, 180–223.

Sawin, P. B. and Edmonds, H. W. (1949). Morphological studies of the rabbit. VII. Aortic arch variations in relation to regionally specific growth differences. *Anat. Rec.*, **96**, 183–200.

Smith, R. J. (1980). Rethinking allometry. *J. Theoret. Biol.*, **87**, 97–111.

Waddington, C. H. (1975). *The Evolution of an Evolutionist*. Edinburgh University Press, Edinburgh.

Chapter 5

Bush, G. L. (1975). Modes of animal speciation. *Ann. Rev. Ecol. Syst.*, **6**, 339–364.

Cronin, J. E., Boaz, N. T., Stringer, C. B. and Rak, Y. (1981). Tempo and mode in hominid evolution. *Nature*, **292**, 113–122.

De Vries, H. (1905). *Species and Varieties. Their Origin by Mutation*. The Open Court, Chicago.

Dobzhansky, T. (1937). *Genetics and the Origin of Species*. Columbia University Press, New York.

Eldredge, N. and Cracraft, J. (1980). *Phylogenetic Patterns and the Evolutionary Process*. Columbia University Press, New York.

Gould, S. J. and Eldredge, N. (1977). Punctuated equilibria: the tempo and mode of evolution reconsidered. *Paleobiology*, **3**, 115–151.

Hennig, W. (1979). *Phylogenetic Systematics*, 2nd edn. University of Illinois Press, Urbana.

Huxley, J. S. (1942). *Evolution, the Modern Synthesis*. Allen & Unwin, London.

Lande, R. (1976). Natural selection and random genetic drift in phenotypic evolution. *Evolution*, **30**, 314–334.

Lewin, R. (1980). Evolution theory under fire. *Science*, **210**, 883–887.

Maynard Smith, J. (1981). Macroevolution. *Nature*, **289**, 13–14.

Mayr, E. (1942). *Systematics and the Origin of Species*. Columbia University Press, New York.

Mayr, E. (1970). *Populations, Species and Evolution*. Harvard University Press, Cambridge, Mass.

Mayr, E. and Provine, W. B. (1980). *The Evolutionary Synthesis*. Harvard University Press, Cambridge, Mass.

Simpson, G. G. (1944). *Tempo and Mode in Evolution*. Columbia University Press, New York.

Sneath, P. H. A. and Sokal, R. R. (1973). *Numerical Taxonomy*. W. H. Freeman, San Francisco.

Sokal, R. R. and Sneath, P. H. A. (1963). *Principles of Numerical Taxonomy*. W. H. Freeman, San Francisco.

Stanley, S. M. (1975). A theory of evolution above the species level. *Proc. Nat. Acad. Sci. U.S.A.*, **72**, 646–650.

Stanley, S. M. (1979). *Macroevolution*. W. H. Freeman, San Francisco.

Stanley, S. M. (1982). *The New Evolutionary Timetable*. Basic Books Inc., New York.

Williams, P. G. (1981). Palaeontological documentation of speciation in Cenozoic molluscs from Turkana Basin. *Nature*, **293**, 437–443.

Index